P9-CBE-647

SCIENCE TALES

'Climate change is a familiar story. What's unusual is the way it's told. *Science Tales* deals with some of the most urgent debates in science using pictures, speech bubbles and comic-strip layouts. Cunningham takes a view on such knotty issues as homeopathy and the MMR vaccine, sorting facts from fiction and presenting complex information in a highly accessible way.'
Observer

'Cunningham's charming artwork complements his concise arguments on climate change, the first moon landing and homeopathy, among other subjects. He consistently champions the scientific method over all forms of quackery, and his stark lines and simple layouts give his comic the feel of a scientific analysis. The artwork is uncluttered, leaving little to distract the reader from the exposition, delivered in stripped-back, staccato prose.'
New Scientist

'Cunningham projects a quietly authoritative voice throughout. As the narrator and host for each chapter he is confrontational yet balanced; unflinching in his condemnation of the irrational and the unsupportable without ever lapsing into belligerence. Artistically, his clean lines and his deceptively simple cartooning style perfectly complement the clarity inherent in the delivery of his carefully considered points. Cunningham manages to deftly précis the salient points of each chapter's discussion in an entertaining, engaging, and sometimes slyly witty way. *Science Tales* manages to be somehow simultaneously both succinct and substantive, and a fierce and intelligent promoter of the scientific process over blind superstition and baseless supposition.'
Broken Frontier

'Cunningham's art has clean lines and a continuity that is often graceful, charming and endearing. He speaks with quiet authority on his subjects, but is careful to cite a whole range of sources and research papers.'
Independent

'It's good to see the arguments presented so well, clearly and concisely. What Cunningham does here is rather brilliantly presented, and customarily classy. As an artist his work is equally at home in the stark black and white of *Psychiatric Tales* or the lush and varied colours in *Science Tales*... *Science Tales* is impressive, Cunningham delivering his message with style, great art, even moments of outright comedy. All in all, we have something else to deliver the message of science and reason, and that's a good, good thing.'

Forbidden Planet

'A lovely book which combines its no-nonsense approach with a funny, pro-science attitude.'

Edzard Ernst, *The Pulse*

'It's a meticulous picking apart of the ridiculous web of half-baked facts and fiction that's often woven around one or two grains of truth, usually completely taken and distorted totally out of context, to prove his case. Anyone who enjoyed Darryl's previous work, *Psychiatric Tales*, will definitely enjoy this. Darryl also employs the same understated clinical yet also slightly comical art style.'

Page 45

'A fantastic nonfiction comic book about science, scepticism and denial. Cunningham has a real gift for making complex subjects simple. If you're a *Mythbusters* fan, admire James Randi, enjoyed Ben Goldacre's *Bad Science*, and care about climate change, you'll enjoy this one. More to the point, if you're trying to discuss these subjects with smart but misguided friends and loved ones, this book might hold the key to real dialogue.'

Cory Doctorow, *Boing Boing*

'He has managed to distil the arguments into a wonderfully clear and concise form... a great primer for those seeking arguments to undermine their *Daily Express*-reading uncle.'

Herald Scotland

'Cunningham is extremely good at explaining the links between bad science and profiteering, both by supposed scientists and by the media. It's clear and straightforward at all times, making complex issues simple, but never simplistic.'

Headline Environment

'Cunningham is admirably erudite... The result is persuasive rhetoric: popular science not overly technical, but communicated clearly and with conviction. Cunningham writes with the courtroom eloquence of the prosecuting barrister, denouncing the accused in capital letters, his words as precise as his drawing style is hard-edged.'

Graphic Medicine

'His style is cartoony and raw, but manages to be full of expression and also very evocative. *Science Tales* deserves a wide audience and even if you haven't tried to read something in comic format before, you'll find this easy to follow on the one hand, and thought-provoking on the other.'

Bradford Telegraph & Argus

'Cunningham never accuses people who are swayed by conspiracy theories or pseudoscience of being evil or stupid, and his tone is polite enough to win hearts and minds, provided they're open minds. *Science Tales* will find its home in classrooms and houses with children, where young people will find it and then prick up their ears anytime an adult mentions "getting an adjustment" or "seeing a homeopath". It will remind them that science is a matter of facts, not politics.'

Comics Alliance

'An eye-catching way to get across the important message that a science-based approach to understanding makes far more sense than one that is evidence-free. Much of modern conventional wisdom is framed by the media, corporations and activist groups to attend to their own agendas. Cunningham draws out the fictions and lays bare the facts.'

Chemistry World

DARRYL CUNNINGHAM

SCIENCE TALES

LIES, HOAXES AND SCAMS

Myriad Editions

First published in 2012 by

Myriad Editions
59 Lansdowne Place
Brighton BN3 1FL, UK

www.myriadeditions.com

This edition published in 2013

1 3 5 7 9 10 8 6 4 2

A CIP catalogue record for this book is available from the British Library.

ISBN: 978-1-908434-36-4

Printed in China on paper sourced from sustainable forests.

CONTENTS:

FOREWORD

FOREWORD:

My childhood was filled with fantastical tales of ghosts and UFOs. I read whatever I could on these subjects, rather credulously believing them, simply because they were written down. I powerfully wanted to believe in something more than the ordinary banal world I saw around me. It was an escape from a humdrum home and school life into something more exciting. Why couldn't there be ghosts? What if the Loch Ness monster existed? I searched the skies in the hope that I might see an alien craft, and, disappointingly, never saw one. I craved the fantastic and the bizarre in order to bring colour to a monochrome world, only to realise, as I gained a few critical thinking skills, that the evidence for these things just didn't exist.

Of course, the universe has amazing and strange qualities anyway for those who care to see them. There's no need to believe in fantasy in order to see the extraordinary in the world, when reality offers up so much that is astonishing. The journey I took from credulous believer in nonsense to a more hard-headed state of mind was a tough one, in which I had to dispense with much I believed. Accepting that you are wrong in the light of new evidence can be a painful process, but is a necessary one. Science, unlike religion, is in a continuous state of revision, depending on evidence. If you want the truth then you must go where the facts take you, however uncomfortable that might make you feel. I've argued strongly for the positions I take in the chapters of this book, but I'd like to think I'd be strong enough to change my mind on any of them if the evidence became available. That's a door that must always be kept open.

I've selected these particular subjects because they were the most prominent hot-button science issues at the time of writing. I had noticed over the past couple of years that, whenever I read a science blog or listened to a science podcast, these subjects would come up as the most controversial, time and again. The level of misunderstanding among much of the general public, not just on the issues themselves but about just how the scientific process works, never fails to amaze me. This book is my small attempt to rectify that problem. The order of the subjects in the book moves from 'personal choice' medical issues, like chiropractic and homeopathy, up to larger issues that concern the environment, such as climate change. The chapter on hydraulic fracturing, or fracking, which was first published in the US edition (*How to Fake a Moon Landing*, Abrams 2013), is now included here in this newly revised edition for the UK.

The whole structure of the book is meant to build up a case for critical thinking and the scientific process itself. This book is pro-science and pro-critical thinking. What it isn't is a book promoting a scientific élite whom we must all follow, sheep-like. It is the scientific process itself I'm promoting here, not the scientific establishment, who are just as capable of being fraudulent, corrupted by politics and money or just plain wrong as any group of humans engaged in any activity. We know the scientific process can be relied on, because, if it couldn't be, the lightbulb wouldn't work when you switched it on, your mobile phone would be a useless brick, and satellites wouldn't be orbiting the planet.

Science isn't a matter of faith or just another point of view. Good science is testable, reproducible, and stands the test of time. What doesn't work in science falls away, and what remains is the truth.

Darryl Cunningham
Yorkshire, November 2012

ELECTROCONVULSIVE THERAPY

I WORKED ON AN ACUTE PSYCHIATRIC WARD FOR YEARS. THE WARD HAD AN ELECTROCONVULSIVE THERAPY UNIT ATTACHED.

THIS UNIT WAS USED TWICE A WEEK. I SAW THE PROCESS IN ACTION.

MY VERY UNSCIENTIFIC OBSERVATION WAS THAT THE THERAPY COULD HAVE EITHER GREAT BENEFIT OR NONE.

IT COULD BRING PEOPLE BACK FROM THE DEPTHS OF SEVERE DEPRESSION...

OR IT COULD HAVE NO EFFECT AT ALL.

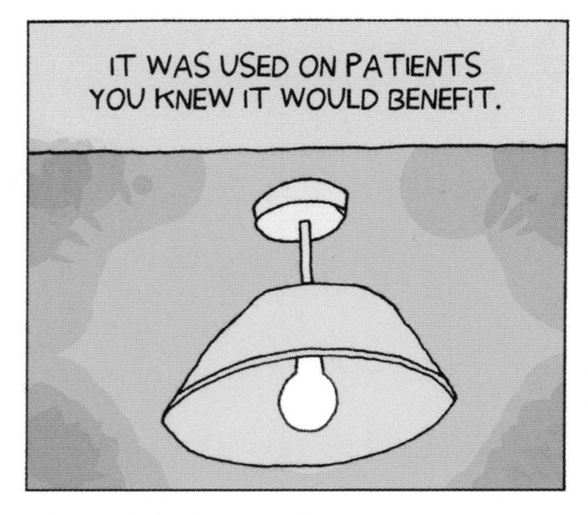
IT WAS USED ON PATIENTS YOU KNEW IT WOULD BENEFIT.

BUT ALSO ON PATIENTS SIMPLY BECAUSE THE PSYCHIATRIST IN CHARGE...

HAD USED UP ALL OTHER OPTIONS AND COULDN'T THINK OF ANYTHING ELSE.

BUT WHEN ECT WORKED IT COULD BRING ABOUT A POSITIVE DIFFERENCE.

ECT TREATMENT INVOLVES PASSING AROUND EIGHT HUNDRED MILLIAMPS OF ELECTRICITY THROUGH THE BRAIN.

SLIGHTLY MORE THAN A MOBILE PHONE BATTERY PUTS OUT.

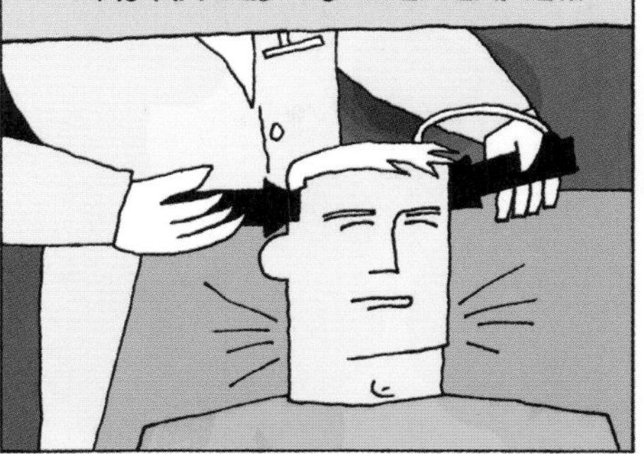
IN THE EARLY DAYS A STIFF JOLT OF HOUSE CURRENT WAS APPLIED TO THE TEMPLE...

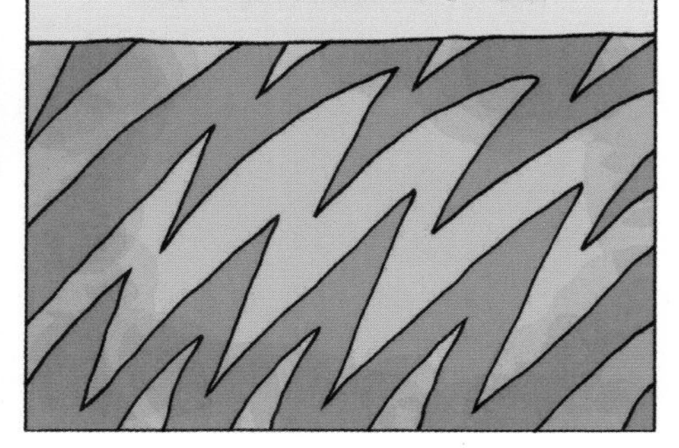
AND THE PATIENT WOULD THEN GO INTO CONVULSIONS LASTING A MINUTE OR SO.

THESE SPASMS WERE OFTEN SO VIOLENT THAT THE PATIENT COULD SUFFER BROKEN BONES OR DISLOCATION.

NOW AN ANAESTHETIC IS USED IN COMBINATION WITH A MUSCLE RELAXANT.

PATIENTS AREN'T CONSCIOUS DURING THE PROCEDURE. NOR DO THEY CONVULSE IN A DANGEROUS MANNER.

THE EFFECT ON THE BRAIN IS ESSENTIALLY THE SAME AS AN EPILEPTIC SEIZURE.

ECT IS USUALLY GIVEN IN DOSES OF SIX TO TWELVE TREATMENTS OVER A SIMILAR PERIOD OF WEEKS...

AFTER WHICH THERE CAN BE A MAINTENANCE TREATMENT DONE LESS FREQUENTLY OVER LONGER PERIODS.

WHY SUCH A THING WOULD WORK AS A BENEFICIAL WAY OF STIMULATING THE BRAIN IS STILL UNKNOWN.

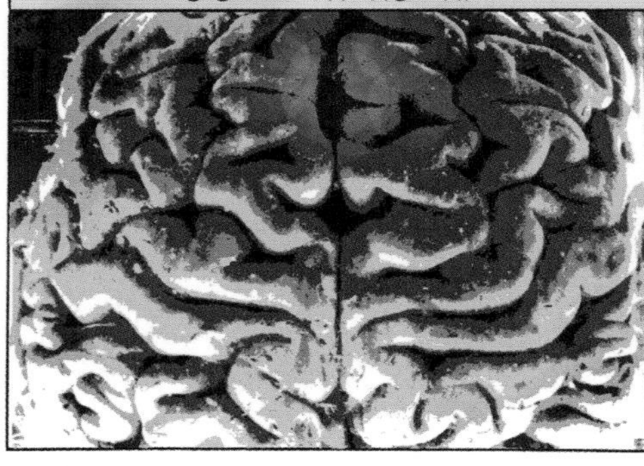

BUT IT REMAINS THE MOST RAPID AND EFFECTIVE TREATMENT FOR SEVERE DEPRESSION.

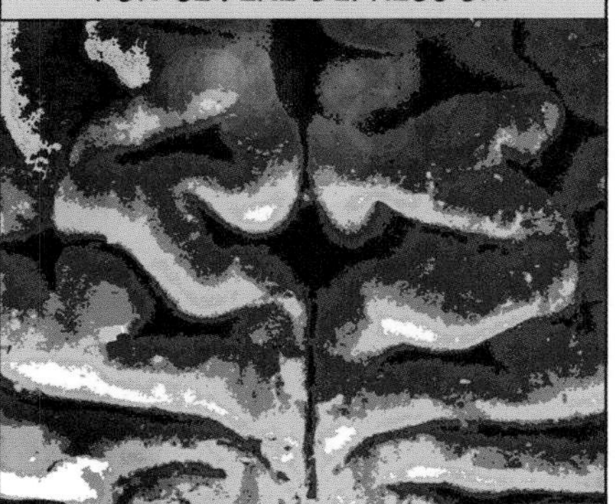

THE NEGATIVES ARE THAT PATIENTS CAN SUFFER MEMORY PROBLEMS WHICH MAY PERSIST FOR SOME TIME.

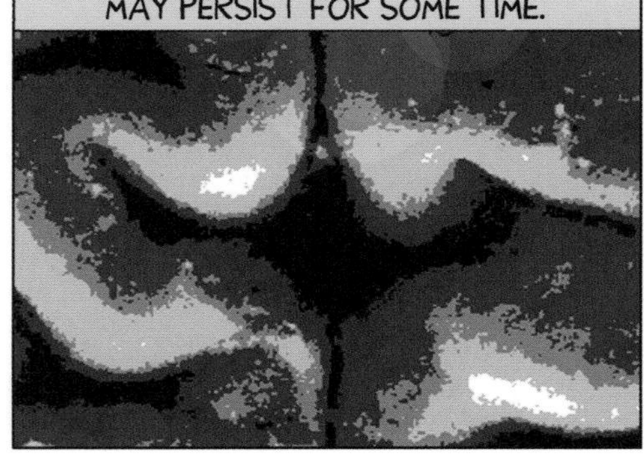

AFTER I WROTE ONLINE THAT I WAS PLANNING TO WRITE A CHAPTER ON ECT, I RECEIVED AN EMAIL FROM AN INDIVIDUAL WHO'D RECENTLY HAD THE TREATMENT. HE SPEAKS HERE IN HIS OWN WORDS.

THANKS, DARRYL. AFTER TEN YEARS STRUGGLING WITH A MOOD DISORDER, I FINALLY CONSENTED TO ECT A FEW MONTHS AGO.

MY FAMILY TELL ME THAT THEY NOTICE QUITE A DIFFERENCE.

BUT I STARTED ON A NEW COCKTAIL OF MEDS AT THE SAME TIME.

SO WHO CAN SAY WHICH COURSE OF TREATMENT MADE THE DIFFERENCE?

I DO FEEL IMPACTED BY THE MEMORY LOSS. I WAS AFRAID THAT IT WOULD MAKE ME DUMB. WELL, IT HASN'T DONE THAT EXACTLY.
D

I CAN STILL UNDERSTAND AND COMPREHEND THINGS, BUT I'M HAVING A HARD TIME COMMITTING THEM TO MEMORY.

WHEN WRITING A GROCERY LIST, I FIND THAT AS SOON AS I WRITE THE FIRST ITEM I'VE FORGOTTEN THE REST OF THE THINGS I WANTED ON THE LIST.

I CAN READ AND ENJOY A BOOK, BUT, AS SOON AS I'VE PUT THAT BOOK DOWN, I FIND THAT I CAN'T REMEMBER ANY OF THE DETAILS.

THE EXPERIENCE OF READING IS STILL ENJOYABLE, BUT I CAN'T RETAIN NEW INFORMATION.

I'M HOPING THAT MY SHORT-TERM MEMORY WILL IMPROVE OVER TIME, BUT I'M AWARE IT MIGHT NOT.

PATIENTS GENERALLY VIEW E C T MUCH LESS FAVOURABLY THAN DO CLINICIANS.

IT'S SEEN EXTREMELY NEGATIVELY BY THE PUBLIC AND CARRIES A SIGNIFICANT STIGMA.

I'VE HAD ELECTRO-CONVULSIVE THERAPY.
!
WHAT?

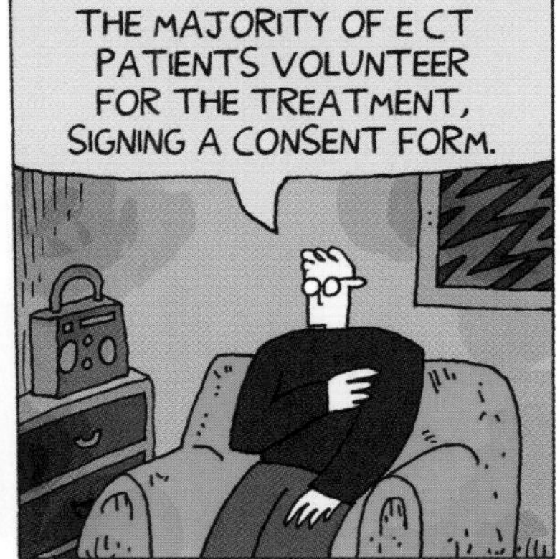
THE MAJORITY OF E C T PATIENTS VOLUNTEER FOR THE TREATMENT, SIGNING A CONSENT FORM.

IN SOME COUNTRIES THE LAW ALLOWS A MINORITY OF PATIENTS TO BE TREATED AGAINST THEIR WILL.
THAT'S HORRIFYING! I DIDN'T KNOW THEY STILL DID THAT.

BUT ONLY WHERE THEY ARE SEEN AS A DANGER TO THEMSELVES OR TO OTHERS.

IT'S REALLY HARD TO HAVE A SENSIBLE DISCUSSION ABOUT ECT BECAUSE OF THE EMOTION IT STIRS UP.

YES, IT DOES SOUND BARBARIC, BUT THE SAME COULD BE SAID OF MANY MEDICAL PROCEDURES.

THE TROUBLE WITH ECT COMPARED TO SURGERY IS THAT IT SOUNDS LIKE SOMETHING FROM ANOTHER AGE.

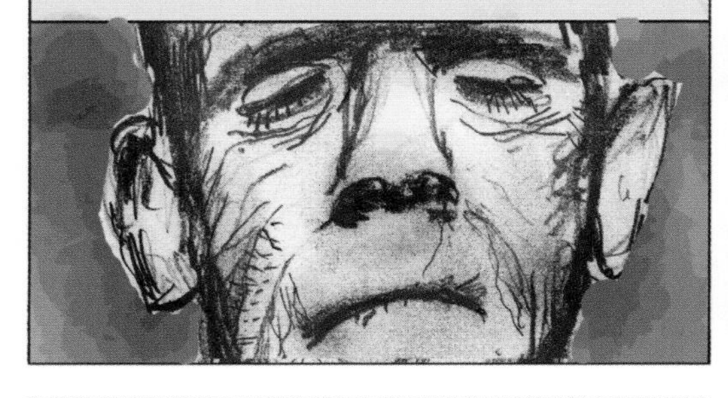

SOMETHING PRIMAL AND MORE AKIN TO ALCHEMY.

IT DOESN'T SEEM TO FIT IN WITH TWENTY-FIRST-CENTURY MEDICAL SCIENCE.

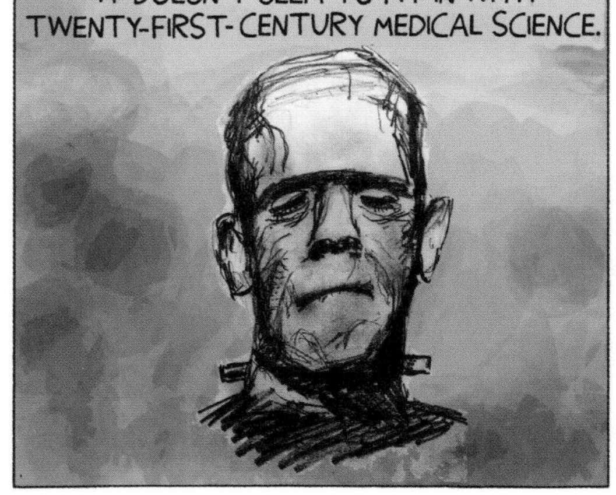

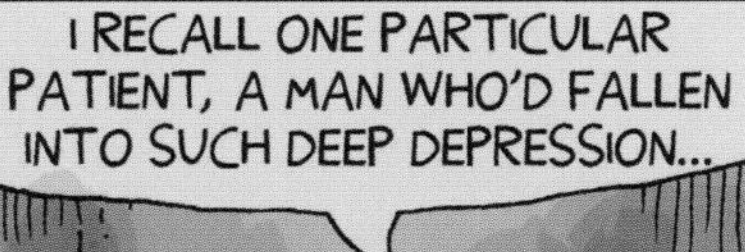

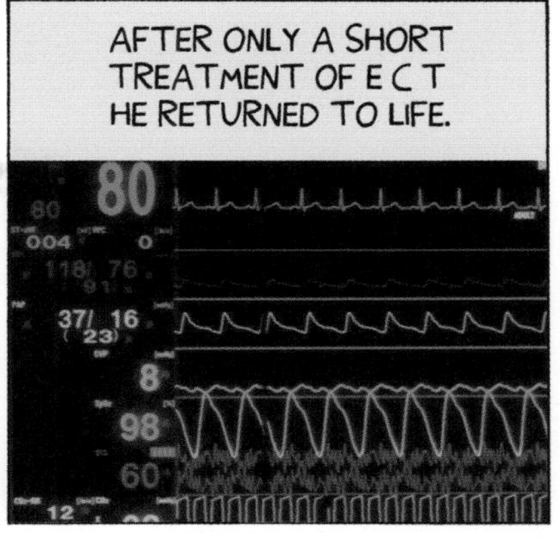

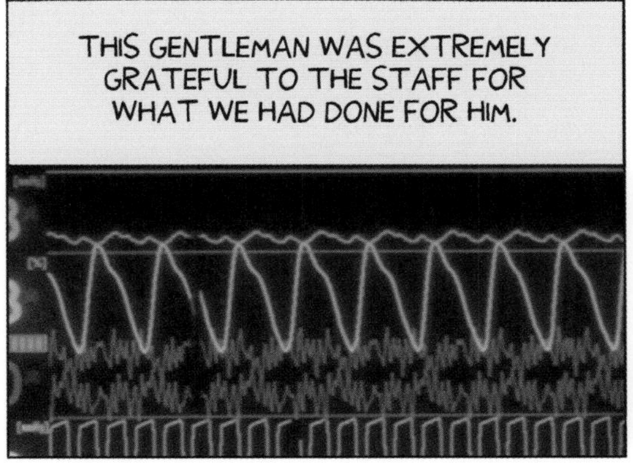

MANY DOCTORS SAY THAT THEY HAVE SEEN ECT RELIEVE SEVERE DEPRESSION WHERE OTHER TREATMENTS HAVE FAILED.

OPPONENTS OF ECT SAY THAT THE TREATMENT IS INHUMANE AND DEGRADING. THAT THE SIDE-EFFECTS ARE SEVERE...

MY PERSONAL VIEW IS THAT IT'S LONG PAST TIME THAT MEDICAL SCIENCE GAVE US SOMETHING MORE EFFECTIVE...

HOMEOPATHY

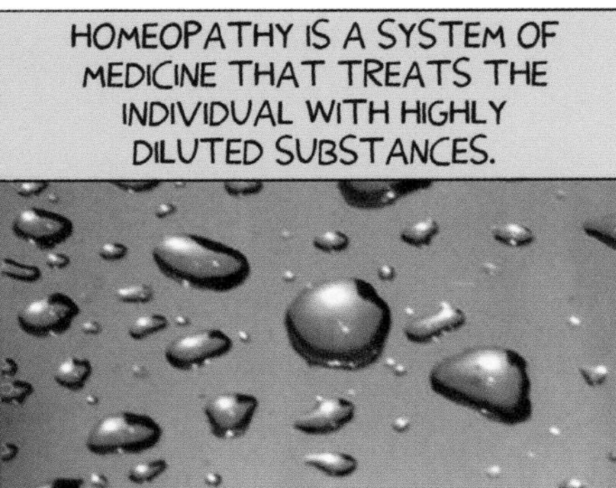
HOMEOPATHY IS A SYSTEM OF MEDICINE THAT TREATS THE INDIVIDUAL WITH HIGHLY DILUTED SUBSTANCES.

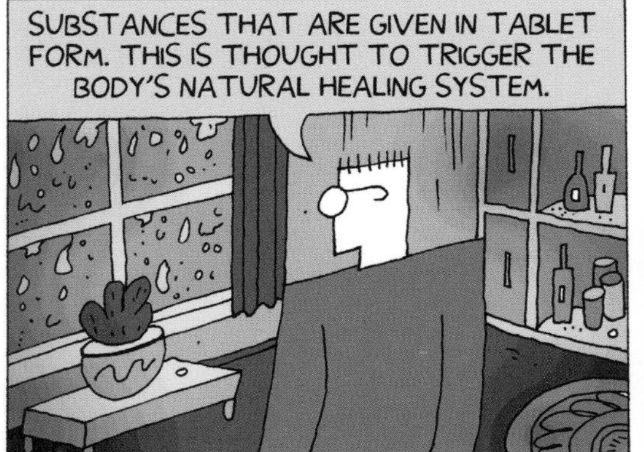
SUBSTANCES THAT ARE GIVEN IN TABLET FORM. THIS IS THOUGHT TO TRIGGER THE BODY'S NATURAL HEALING SYSTEM.

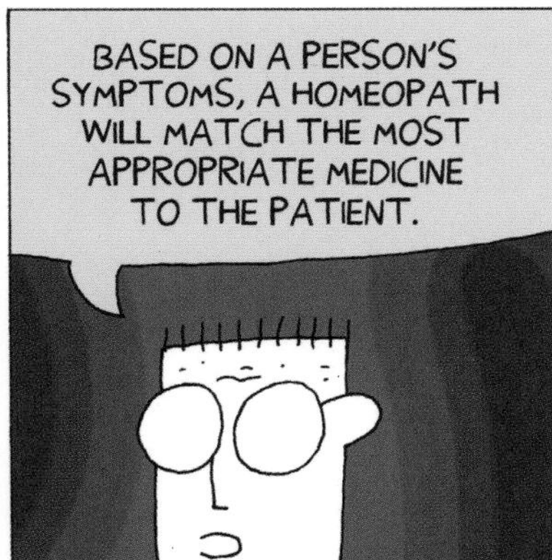
BASED ON A PERSON'S SYMPTOMS, A HOMEOPATH WILL MATCH THE MOST APPROPRIATE MEDICINE TO THE PATIENT.

HOMEOPATHY IS BASED ON TWO MAIN HYPOTHESES. THE FIRST IS THE LAW OF SIMILARS...

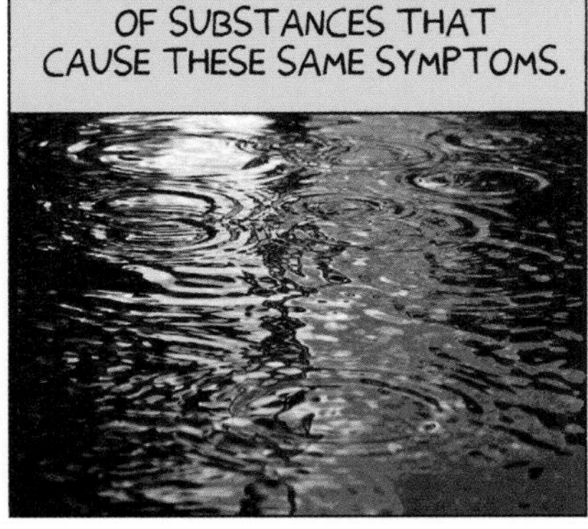
THE IDEA THAT ILLNESSES CAN BE CURED BY SMALL DOSES OF SUBSTANCES THAT CAUSE THESE SAME SYMPTOMS.

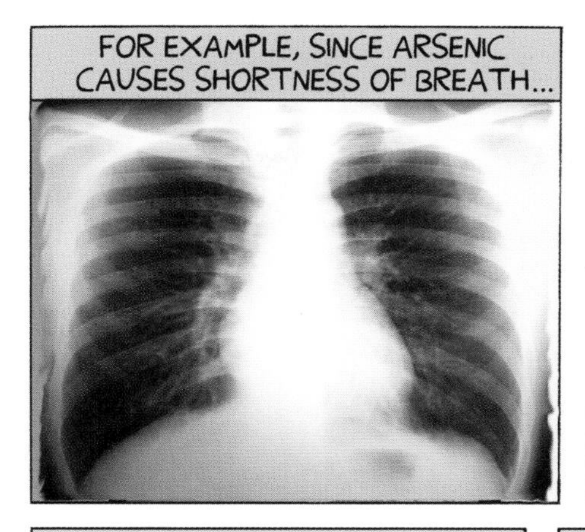
FOR EXAMPLE, SINCE ARSENIC CAUSES SHORTNESS OF BREATH...

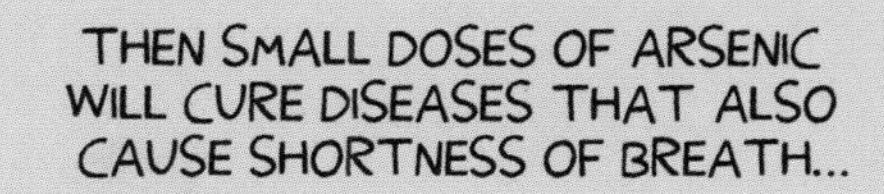
THEN SMALL DOSES OF ARSENIC WILL CURE DISEASES THAT ALSO CAUSE SHORTNESS OF BREATH...

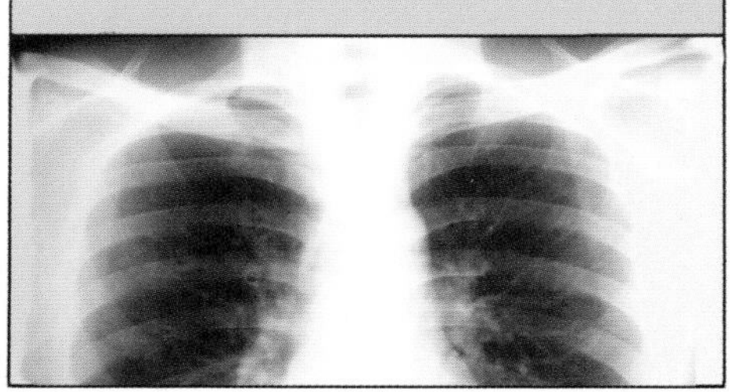

SUCH AS ASTHMA.

THE SECOND IDEA IS THE SMALLER THE DOSE, THE STRONGER THE CURE.

THE ACTIVE INGREDIENT, LET'S SAY ARSENIC IN THIS CASE, IS THEN REPEATEDLY DILUTED.

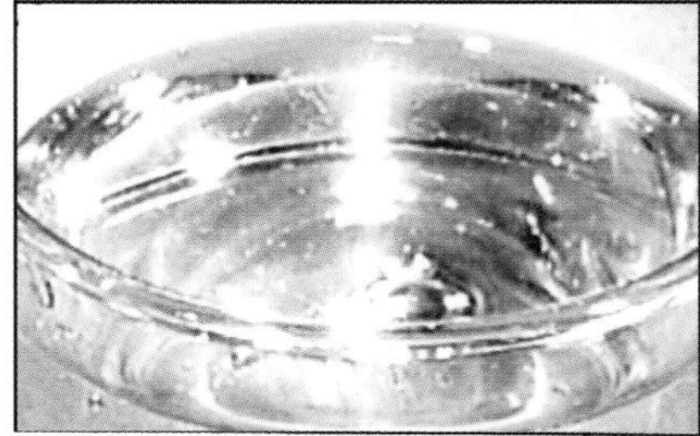
THIS PROCESS OF DILUTION CONTINUES UNTIL IT'S HIGHLY UNLIKELY THAT EVEN A SINGLE MOLECULE OF THE ACTIVE INGREDIENT REMAINS IN THE POTION.

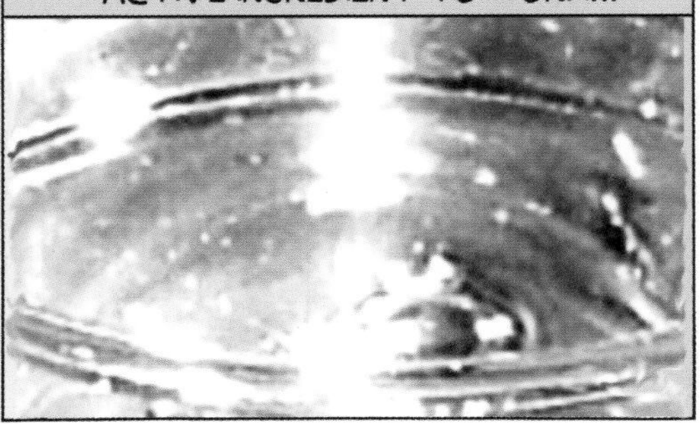
HOWEVER, ACCORDING TO HOMEOPATHIC THEORY, THE DILUTED REMEDY NEEDS NO TRACE OF THE ACTIVE INGREDIENT TO WORK...

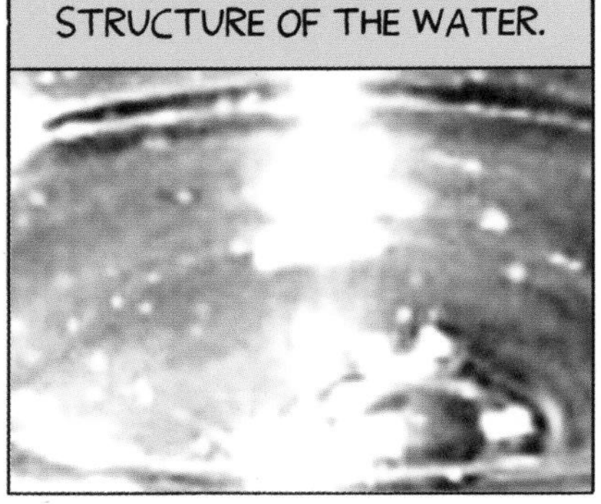
AS THE SUBSTANCE IS IMPRINTED ON THE STRUCTURE OF THE WATER.

EACH TIME THE INGREDIENT IS DILUTED, IT IS VIGOROUSLY SHAKEN.

A PROCESS THAT COMPELS THE WATER TO 'REMEMBER' THE GRADUALLY VANISHING INGREDIENT.

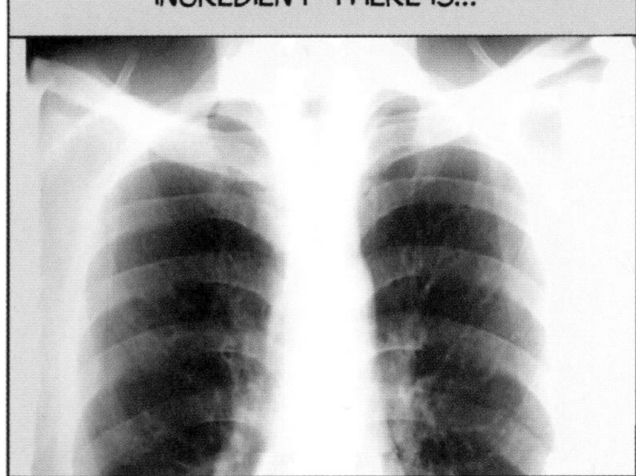
THE MORE THE REMEDY IS SHAKEN, AND THE LESS OF THE ORIGINAL INGREDIENT THERE IS...

THE MORE POTENT AND POWERFUL THE REMEDY BECOMES.

CAN WATER HAVE A MEMORY?

IN 1988, JACQUES BENVENISTE, A FRENCH BIOLOGIST, CLAIMED THAT IT COULD.

THE HIGHLY RESPECTED JOURNAL NATURE PUBLISHED HIS CONTROVERSIAL RESEARCH PAPER.
nature

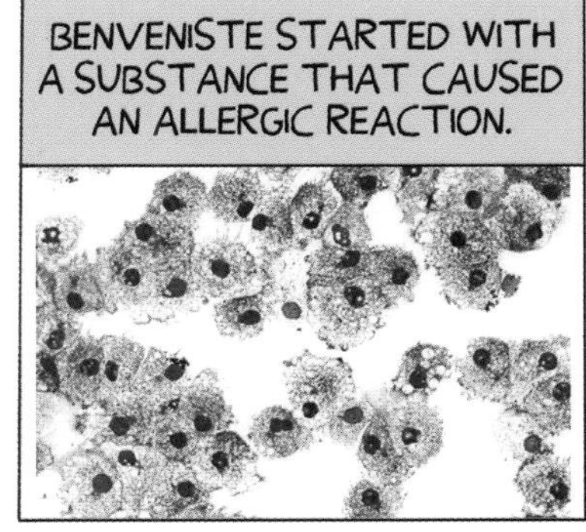
BENVENISTE STARTED WITH A SUBSTANCE THAT CAUSED AN ALLERGIC REACTION.

THIS WAS DILUTED REPEATEDLY UNTIL THERE WAS NOTHING LEFT EXCEPT PURE WATER.

YET THIS WATER STILL MANAGED TO TRIGGER AN ALLERGIC REACTION WHEN ADDED TO LIVING CELLS.

JOHN MADDOX, EDITOR OF NATURE, REALISED THAT BENVENISTE'S RESEARCH WOULD BE CONTROVERSIAL.

THE INVESTIGATION TEAM WAS LED BY MADDOX HIMSELF. HE WAS JOINED BY CHEMIST WALTER STEWART...

AND JAMES RANDI, A MAGICIAN, KNOWN FOR HIS EXPERTISE IN INVESTIGATING EXTRAORDINARY CLAIMS.

WHEN THE BENVENISTE TEAM REPEATED THE EXPERIMENT, THE INVESTIGATORS WENT TO EXTREME LENGTHS TO ENSURE THAT NONE OF THE SCIENTISTS KNEW WHICH SAMPLES...

WERE HOMEOPATHIC SOLUTIONS, AND WHICH WERE THE CONTROLS.

EVEN GOING SO FAR AS TO TAPE SAMPLES TO THE CEILING FOR THE DURATION.

THE INVESTIGATORS SOON DISCOVERED THAT THE RESULTS IN BENVENISTE'S LABORATORY WERE UNRELIABLE.

THEY FOUND A MIXTURE OF LOOSE OR NON-EXISTENT CONTROLS, POSSIBLE EQUIPMENT CONTAMINATION, DATA MANIPULATION, AND DATA SELECTION (SELECTING FOR POSITIVE RESULTS AND IGNORING THE NEGATIVE).

WE BELIEVE THAT EXPERIMENTAL DATA HAVE BEEN UNCRITICALLY ASSESSED AND THEIR IMPERFECTIONS INADEQUATELY REPORTED.

DESPITE NATURE MAGAZINE'S DAMNING REPORT, BENVENISTE CONTINUED TO MAINTAIN THAT HIS RESEARCH WAS VALID.

HE LATER FOUNDED A COMPANY, CALLED DIGIBIO, WHICH MADE THE CLAIM NOT ONLY THAT WATER HAD A MEMORY...

BUT THAT THIS MEMORY COULD BE DIGITISED, TRANSMITTED VIA EMAIL, AND THEN REINTRODUCED INTO WATER.

THERE HAVE BEEN MANY ATTEMPTS TO REPRODUCE BENVENISTE'S EXPERIMENTS. BUT ANY POSITIVE RESULTS HAVE BEEN NEITHER CONSISTENT NOR CONVINCING.

ALL EVIDENCE POINTS TO HOMEOPATHIC REMEDIES BEING INERT AND NO MORE EFFECTIVE THAN A PLACEBO...

OR JUST LETTING THE ILLNESS RUN ITS COURSE.

ARE THERE ANY ASPECTS OF HOMEOPATHY THAT CAN BE SAID TO WORK?
READER'S VOICE

YES. A HOMEOPATH WILL SPEND AT LEAST AN HOUR, SOMETIMES LONGER...

ASKING DETAILED QUESTIONS ABOUT THE PATIENT'S CURRENT HEALTH, MEDICAL HISTORY, AND LIFESTYLE.

IT'S KNOWN THAT STRESS AND ANXIETY CAN ENHANCE ILLNESS.

SO ANYTHING THAT DIMINISHES STRESS AND ANXIETY, SUCH AS THE ATTENTION A HOMEOPATH GIVES...
?

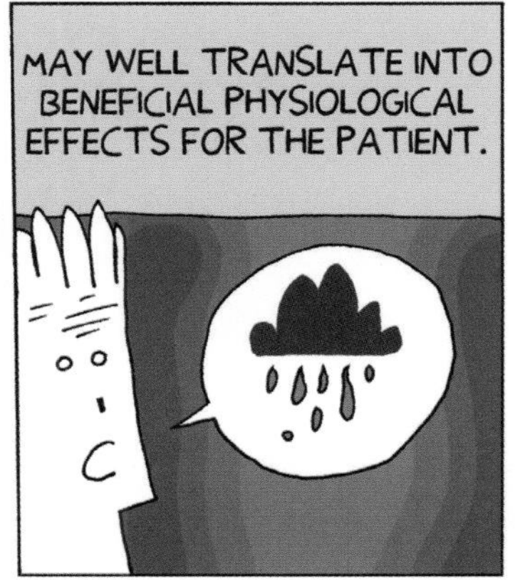
MAY WELL TRANSLATE INTO BENEFICIAL PHYSIOLOGICAL EFFECTS FOR THE PATIENT.

PATIENTS OF HOMEOPATHY FEEL THAT THEY ARE BEING TREATED AS AN INDIVIDUAL.

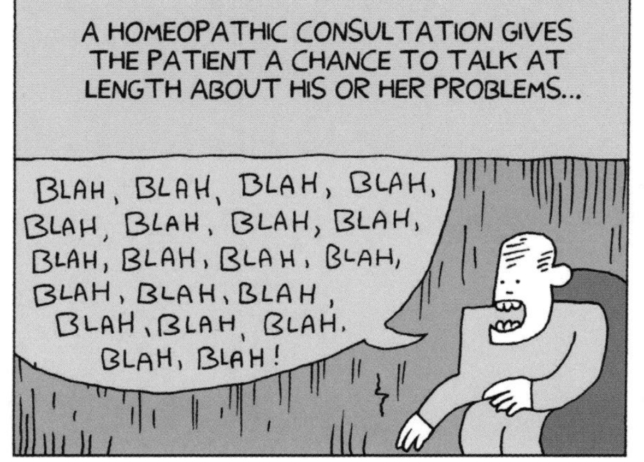
A HOMEOPATHIC CONSULTATION GIVES THE PATIENT A CHANCE TO TALK AT LENGTH ABOUT HIS OR HER PROBLEMS...
BLAH, BLAH, BLAH, BLAH, BLAH, BLAH, BLAH, BLAH, BLAH, BLAH, BLAH, BLAH, BLAH, BLAH, BLAH, BLAH, BLAH, BLAH. BLAH, BLAH!

TO A SYMPATHETIC LISTENER, IN A STRUCTURED ENVIRONMENT.
DO GO ON.

THIS IN ITSELF IS THERAPEUTIC. SO, TO ANSWER YOUR QUESTION, THE ASPECT OF HOMEOPATHY THAT DOES WORK...

IS ACTUALLY A FORM OF PSYCHOTHERAPY.
I SEE.
READER'S VOICE

WHAT THEN IS THE HARM IN HOMEOPATHY IF THE REMEDIES ARE INERT...

AND THE CONSULTATIONS HAVE A POSITIVE THERAPEUTIC EFFECT?

SURELY THERE CAN'T BE ANY DANGERS ATTACHED TO THE PRACTICE?

THE MAIN DANGER OF HOMEOPATHY IS THAT IT CAN DISCOURAGE PEOPLE FROM GETTING REAL TREATMENT.

HOMEOPATHY IS NEVER GOING TO CURE A YEAST INFECTION, ASTHMA, OR CANCER.

AN AVOIDANCE OF SCIENCE-BASED MEDICINE CAN ONLY LEAD TO SICKNESS...

AND DEATH.

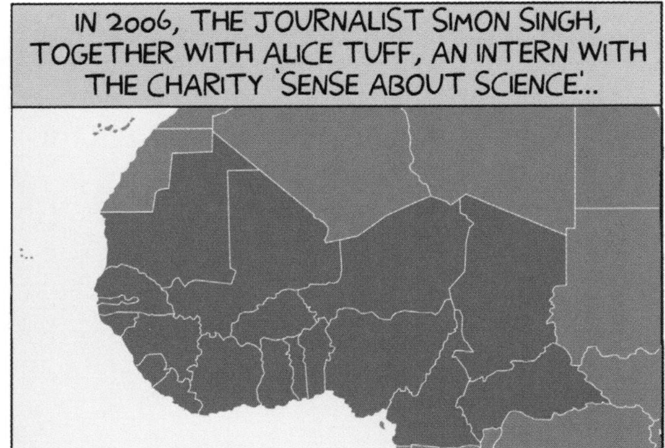
IN 2006, THE JOURNALIST SIMON SINGH, TOGETHER WITH ALICE TUFF, AN INTERN WITH THE CHARITY 'SENSE ABOUT SCIENCE'...

INVESTIGATED THIS ISSUE. TUFF TOOK THE ROLE OF A YOUNG STUDENT, WHO CLAIMED TO BE ABOUT TO MAKE A TRIP THROUGH WEST AFRICA...

WHERE MALARIA IS AT ITS MOST DEADLY.

TUFF CONTACTED TEN HOMEOPATHY CLINICS AROUND LONDON AND ASKED FOR ADVICE.

ALL TEN HOMEOPATHS ADVISED HOMEOPATHIC PROTECTION AGAINST MALARIA.

NONE OF THEM RECOMMENDED TAKING BOTH THE CONVENTIONAL TREATMENT AND THE HOMEOPATHIC REMEDIES.

VERY DANGEROUS FOR TUFF, IF SHE HAD ACTUALLY INTENDED TO GO TO WEST AFRICA.

IN 2005, THE HEALTH PROTECTION AGENCY, AN INDEPENDENT UK ORGANISATION, ISSUED A WARNING...
Health
Protection
Agency

DUE TO THE NUMBER OF PEOPLE FALLING ILL AFTER USING HOMEOPATHIC REMEDIES.

THERE IS NO SCIENTIFIC PROOF THAT HOMEOPATHIC REMEDIES ARE EFFECTIVE IN PREVENTING OR TREATING MALARIA.
SPOKES-MAN

THERE ARE ALSO DANGERS WHEN HOMEOPATHS REPLACE DOCTORS AS A SOURCE OF MEDICAL ADVICE.

IN 2002, EDZARD ERNST AND KATJA SCHMIDT OF THE UNIVERSIY OF EXETER...

CONDUCTED A REVEALING SURVEY AMONG UK HOMEOPATHS.

THEY SENT AN EMAIL TO 168 HOMEOPATHS, IN WHICH THEY PRETENDED TO BE A MOTHER...
QUACK! QUACK!

ASKING FOR ADVICE ON WHETHER SHE SHOULD VACCINATE HER CHILD AGAINST MEASLES, MUMPS AND RUBELLA.

OF THE 77 THAT REPLIED, ONLY TWO ADVISED THAT THE MOTHER SHOULD IMMUNISE.

ANY CHILD NOT IMMUNISED WOULD BE AT RISK.

STORIES OF PEOPLE ABANDONING REAL MEDICINE IN FAVOUR OF QUACK CURES ARE NOT HARD TO FIND.

HERE'S JUST ONE.

PENELOPE DINGLE OF PERTH, AUSTRALIA, WAS DIAGNOSED WITH COLON CANCER IN 2003. HER DOCTORS GAVE HER A GOOD CHANCE OF SURVIVAL WITH STANDARD THERAPY.

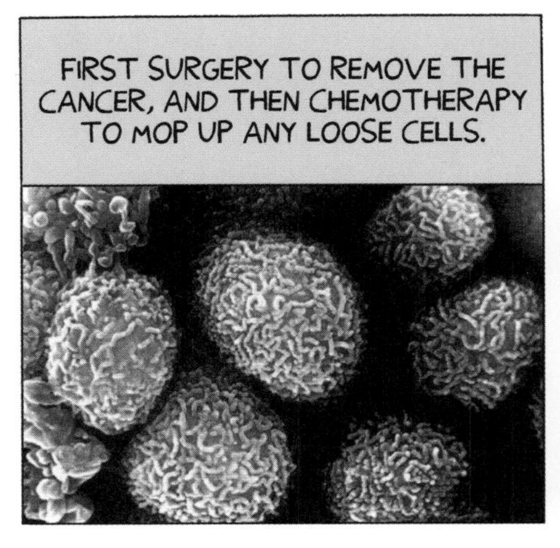
FIRST SURGERY TO REMOVE THE CANCER, AND THEN CHEMOTHERAPY TO MOP UP ANY LOOSE CELLS.

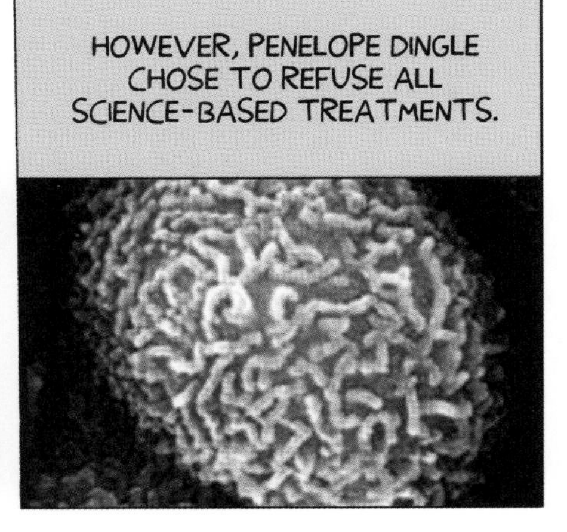
HOWEVER, PENELOPE DINGLE CHOSE TO REFUSE ALL SCIENCE-BASED TREATMENTS.

SHE TRUSTED INSTEAD IN HOMEOPATHY AND NUTRITIONAL SUPPLEMENTS.

HER DIARIES FROM THIS TIME SHOW THAT SHE SAW THE CANCER AS A TEST OF HER FAITH IN ALTERNATIVE MEDICINE.

IN OCTOBER 2003, DEBORAH COMBES, A REGISTERED NURSE AND FAMILY FRIEND,
PEN DINGLE

WAS ASKED BY PEN DINGLE'S SISTERS TO GO AND HELP.

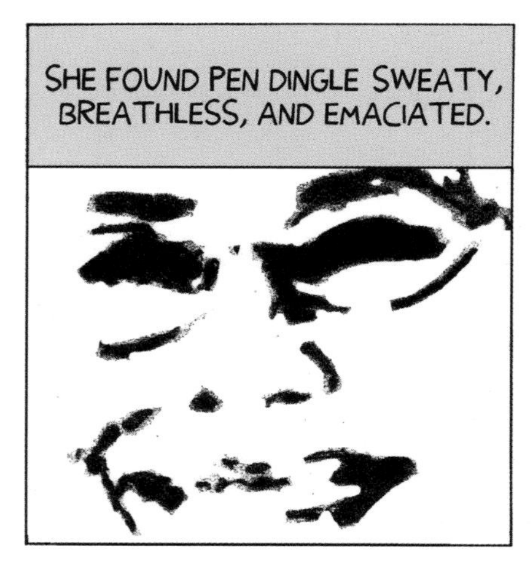
SHE FOUND PEN DINGLE SWEATY, BREATHLESS, AND EMACIATED.

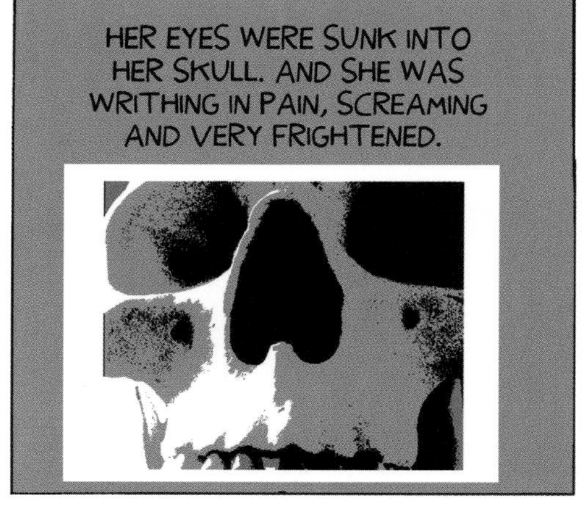
HER EYES WERE SUNK INTO HER SKULL. AND SHE WAS WRITHING IN PAIN, SCREAMING AND VERY FRIGHTENED.

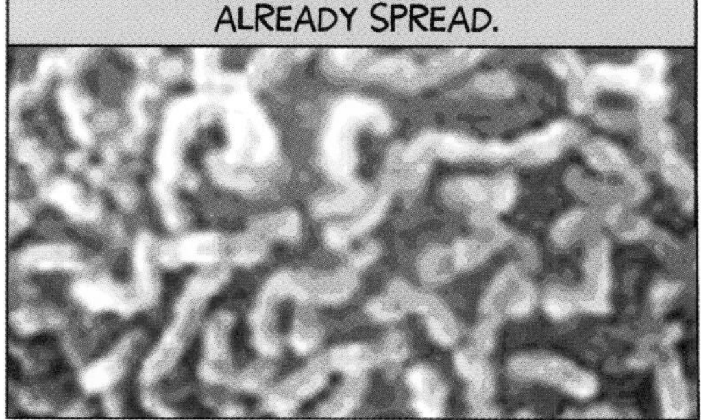
SHE WAS RUSHED INTO HOSPITAL AND HAD THE TUMOUR REMOVED. HOWEVER, THE CANCER HAD ALREADY SPREAD.

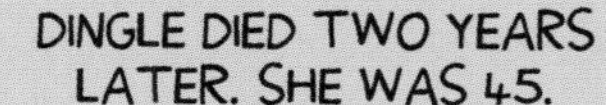
DINGLE DIED TWO YEARS LATER. SHE WAS 45.

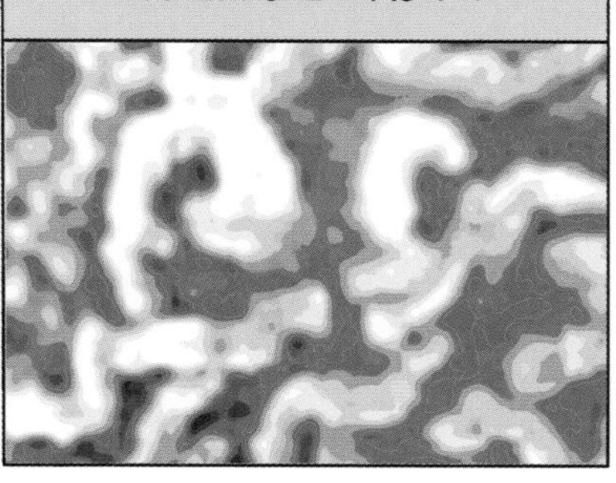

PENELOPE DINGLE'S HUSBAND...

DR. PETER DINGLE, AN ASSOCIATE PROFESSOR IN HEALTH AND ENVIRONMENT...

IS SOMETHING OF A MEDIA CELEBRITY IN AUSTRALIA, AN EXPERT ON DIET AND NUTRITION, LIFESTYLE, AND ENVIRONMENTAL HEALTH.

HE HAS A BACHELOR OF EDUCATION IN SCIENCE, A BACHELOR OF ENVIRONMENTAL SCIENCE WITH FIRST CLASS HONOURS AND A PHD.

DR. DINGLE HAS PERHAPS BEEN UNFAIRLY PORTRAYED IN THE MEDIA, AS A MAN WHO WAS WILLING TO PLACE HIS WIFE AT RISK IN ORDER TO PUT HIS NUTRITIONAL THEORIES TO THE TEST...
ACTUAL BOOK
MY DOG EATS BETTER THAN YOUR KIDS. DR. DINGLE

AND FURTHER HIS OWN CAREER.
MY DOG EATS BETTER THAN YOUR KIDS. DR. DINGLE

HE HAS REPEATEDLY DENIED HAVING ANY REAL INFLUENCE OVER HIS WIFE.

HOWEVER, THERE HAVE BEEN ACCUSATIONS FROM PENELOPE DINGLE'S SISTERS...

THAT DR. DINGLE WAS PLANNING A BOOK IN COLLABORATION WITH PEN'S HOMEOPATH, FRANCINE SCRAYEN...

WHICH WOULD HAVE BEEN ABOUT THE CURING OF PEN'S CANCER USING ONLY ALTERNATIVE MEDICINE AND NUTRITION.

THE CORONER OBSERVED THAT MRS SCRAYEN WAS NOT A COMPETENT HEALTH PROFESSIONAL AND THAT SHE HAD ONLY A MINIMAL UNDERSTANDING OF RELEVANT HEALTH ISSUES. UNFORTUNATELY THIS DID NOT STOP HER FROM TREATING THE DECEASED AS A PATIENT.

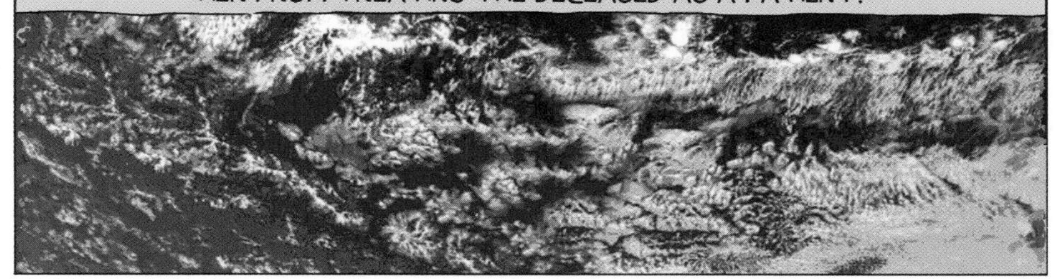

THE CORONER ALSO NOTED THAT THE RELATIONSHIP BETWEEN THE DECEASED AND MRS SCRAYEN WENT FAR BEYOND THE NORMAL PATIENT AND HEALTH PROVIDER RELATIONSHIP.

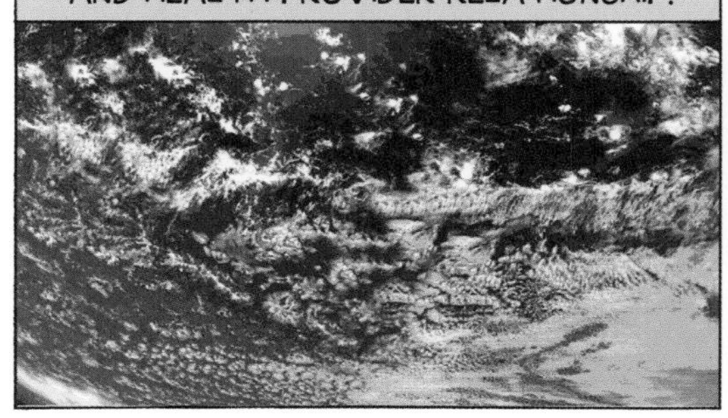

PENELOPE DINGLE HAD BECOME EXTREMELY DEPENDENT ON MRS SCRAYEN.

IN ORDER FOR HOMEOPATHY TO BE EFFECTIVE IT WOULD HAVE TO WORK IN VIOLATION OF THE PRINCIPLES OF BIOLOGY, CHEMISTRY, AND PHYSICS.

HOMEOPATHY HAS NOTHING TO WITH SCIENCE AND EVERYTHING TO WITH FAITH. ITS DEVOTEES DON'T CONCERN THEMSELVES WITH EVIDENCE.

THAT SCIENCE DOESN'T RECOGNISE HOMEOPATHY IS, TO ITS SUPPORTERS, VERY MUCH IN ITS FAVOUR.

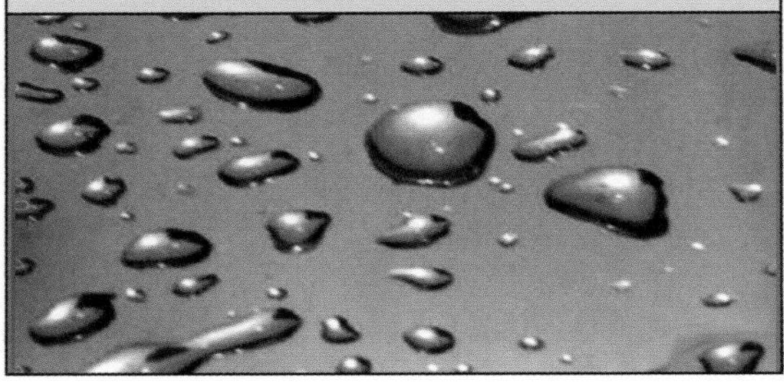

PENELOPE DINGLE BELIEVED IN THE POWER OF HOMEOPATHY.

SHE PUT HER LIFE AT RISK IN ORDER TO PROVE THAT WESTERN MEDICINE WAS UNNECESSARY.

AND THEN SHE DIED.

THE FACTS IN THE CASE OF DR. ANDREW WAKEFIELD

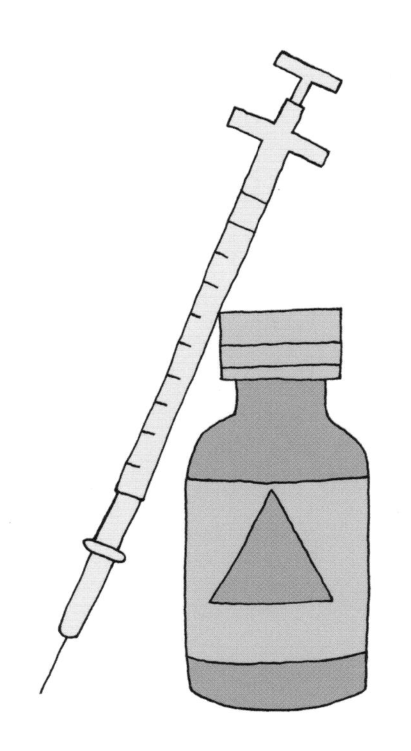

IT'S THE TWENTY-FIRST CENTURY.

AND WE'RE A LONG WAY FROM THE PRE-ENLIGHTENMENT MIDDLE AGES.

THE WORLD HAS BEEN TRANSFORMED BY SCIENTIFIC KNOWLEDGE.

YET SUSPICION OF SCIENCE SEEMS NEVER TO HAVE BEEN HIGHER.

FEAR AND ANGER HAVE OBLITERATED RATIONAL DISCOURSE.

FACTS AND EVIDENCE ARE SEEN AS JUST A MATTER OF OPINION, RATHER THAN A PROVEN TRUTH.

AND BLIND, UNREASONING BELIEF IS CONSIDERED AS VALID AS CRITICAL THINKING.
NO MMR
WE'RE WITH WAKEFIELD
MMR. A JAB TOO FAR

THIS IS ANDREW WAKEFIELD.

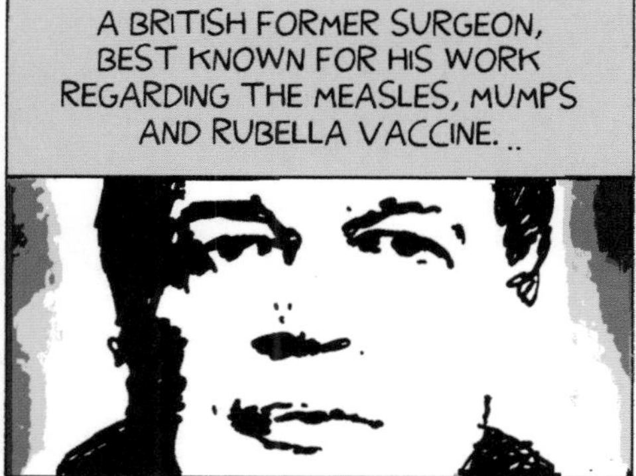
A BRITISH FORMER SURGEON, BEST KNOWN FOR HIS WORK REGARDING THE MEASLES, MUMPS AND RUBELLA VACCINE. ..

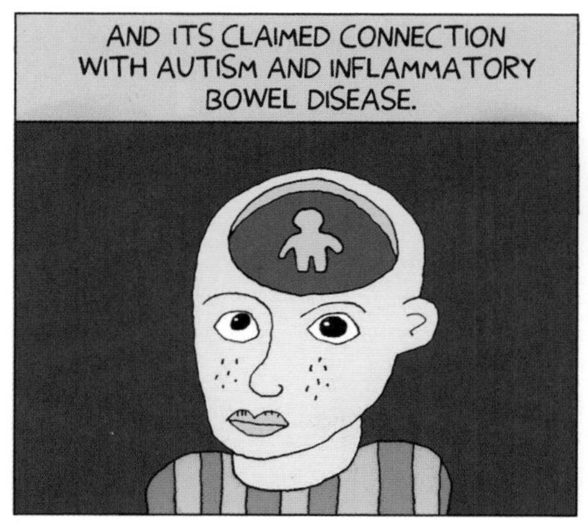
AND ITS CLAIMED CONNECTION WITH AUTISM AND INFLAMMATORY BOWEL DISEASE.

WAKEFIELD WAS THE LEAD AUTHOR IN A 1998 PAPER, PUBLISHED IN THE LANCET.
THE LANCET

THE PAPER REPORTED A STUDY OF TWELVE CHILDREN ALL DIAGNOSED WITH AUTISM...

IN WHICH THE AUTHORS SUGGESTED A LINK WITH THE MMR VACCINE.

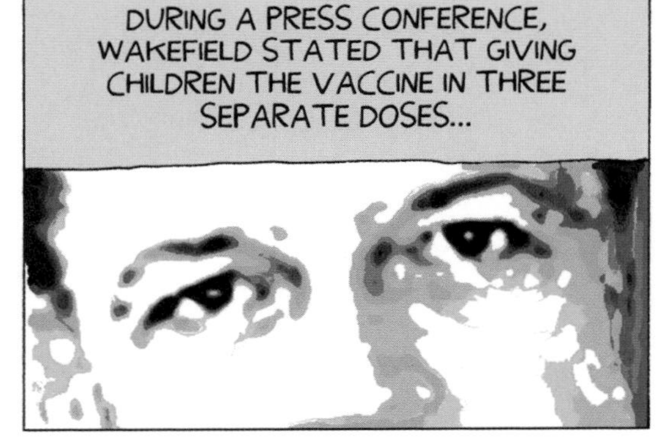
DURING A PRESS CONFERENCE, WAKEFIELD STATED THAT GIVING CHILDREN THE VACCINE IN THREE SEPARATE DOSES...

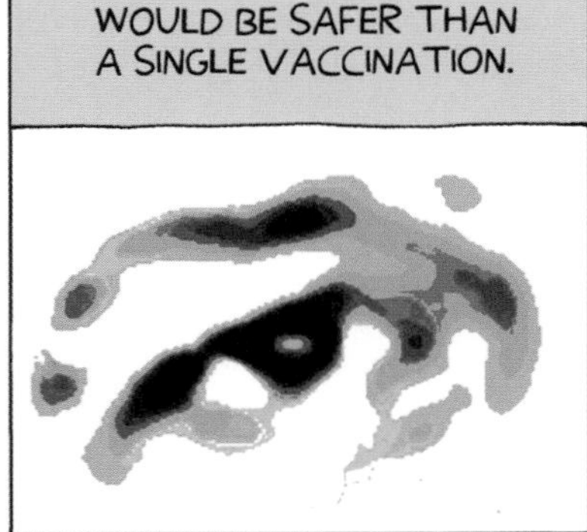
WOULD BE SAFER THAN A SINGLE VACCINATION.

THIS SUGGESTION WAS NOT SUPPORTED BY THE PAPER, AND SUBSEQUENT PEER REVIEW STUDIES...

HAVE NOT SHOWN ANY ASSOCIATION BETWEEN THE VACCINE AND AUTISM.

THIS BEGAN A GLOBAL HEALTH SCARE.

FEAR SPREAD AMONG PARENTS, WHO WERE UNSURE WHAT IMMUNISATION CHOICES TO MAKE.

AND PARENTS OF AUTISTIC CHILDREN BEGAN TO QUESTION THE MMR VACCINATION.
HA! HA!

HE WAS FINE BEFORE THAT AWFUL JAB.

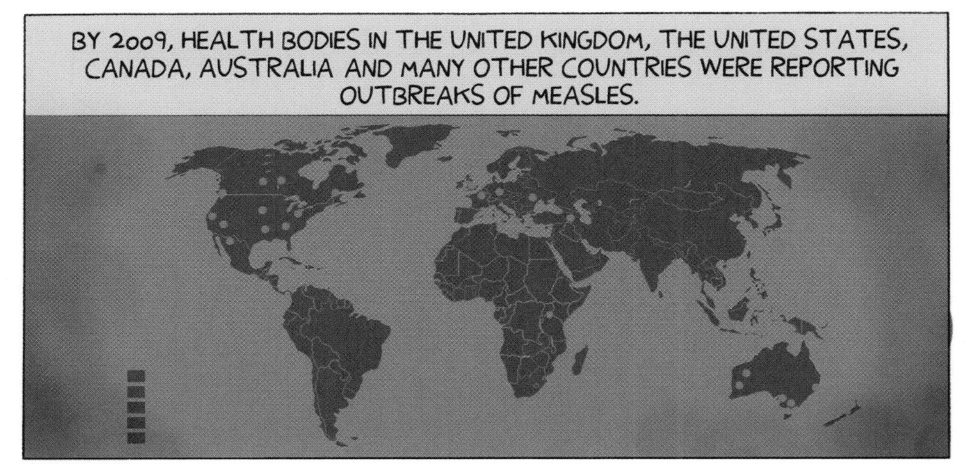
BY 2009, HEALTH BODIES IN THE UNITED KINGDOM, THE UNITED STATES, CANADA, AUSTRALIA AND MANY OTHER COUNTRIES WERE REPORTING OUTBREAKS OF MEASLES.

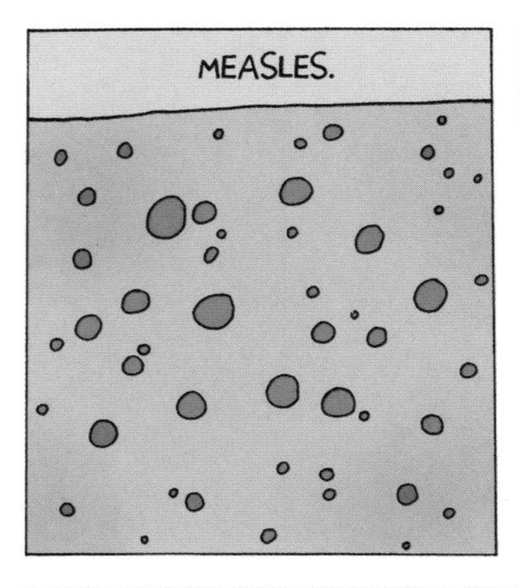
MEASLES.

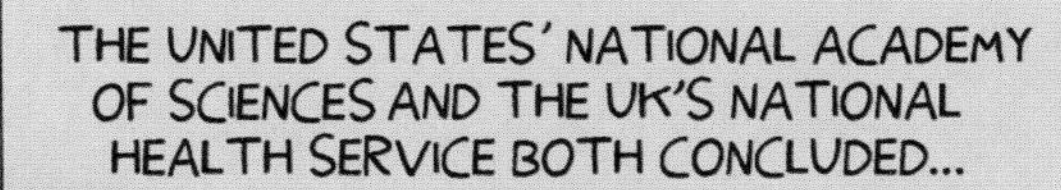
THE UNITED STATES' NATIONAL ACADEMY OF SCIENCES AND THE UK'S NATIONAL HEALTH SERVICE BOTH CONCLUDED...

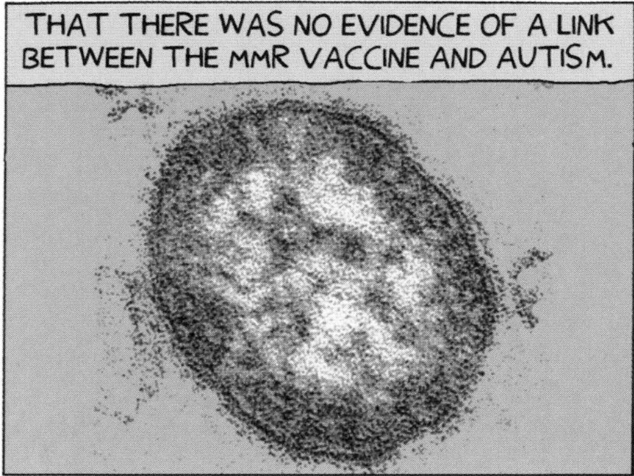
THAT THERE WAS NO EVIDENCE OF A LINK BETWEEN THE MMR VACCINE AND AUTISM.

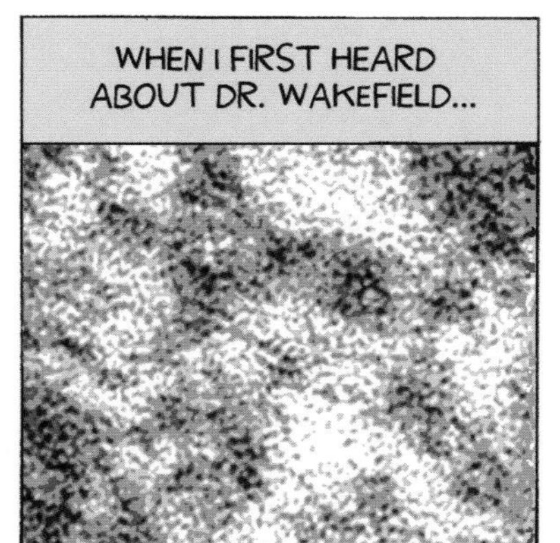
WHEN I FIRST HEARD ABOUT DR. WAKEFIELD...

I THOUGHT, WELL, HE'S A POOR SCIENTIST IF HE CAN DRAW SUCH HUGE CONCLUSIONS FROM SUCH A SMALL STUDY.

AND THERE WAS SOMETHING CLEARLY AMISS IN THE FACT THAT HIS RESULTS HAD FAILED TO BE REPLICATED BY ANYONE ELSE.

IT NEVER OCCURRED TO ME THAT WAKEFIELD MIGHT BE A MAN WHO WAS BENEFITING FINANCIALLY FROM THE POISONED ATMOSPHERE...

OF FEAR, GUILT AND INFECTIOUS DISEASE THAT HE, IN PART, HAD CREATED.

BUT SUCH WAS THE CASE, AND THIS WAS PROVED BY THE JOURNALIST BRIAN DEER...
4

REPORTED FIRST IN THE SUNDAY TIMES AND THEN EXPANDED ON IN A CHANNEL FOUR DOCUMENTARY.
4

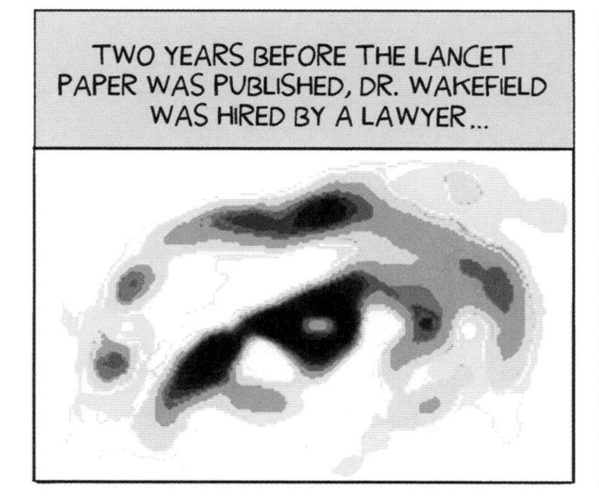
TWO YEARS BEFORE THE LANCET PAPER WAS PUBLISHED, DR. WAKEFIELD WAS HIRED BY A LAWYER...

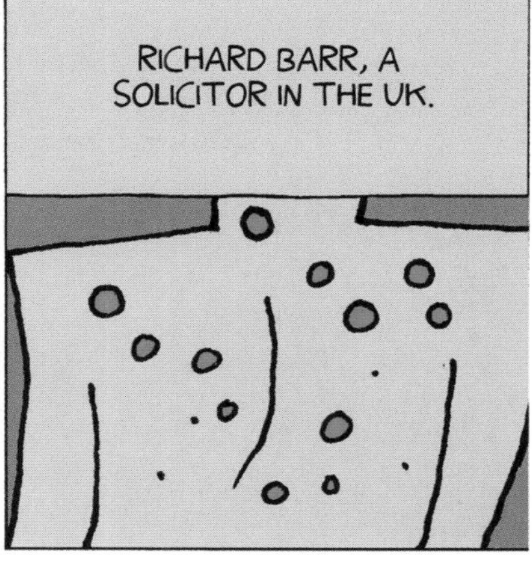
RICHARD BARR, A SOLICITOR IN THE UK.

BARR HOPED TO RAISE A CLASS ACTION LAWSUIT AGAINST MMR MANUFACTURERS.

AND WAKEFIELD WAS CONTRACTED TO CONDUCT SCIENTIFIC RESEARCH FOR HIM.

THE AIM WAS TO FIND EVIDENCE THAT THE MMR VACCINE DID HARM.

THIS EVIDENCE WAS INTENDED TO FEATURE IN LITIGATION ON BEHALF OF 1,600 FAMILIES.

FOR THIS WORK, THE DOCTOR WAS PAID A STAGGERING £150 AN HOUR...

WHICH WAS PAID OUT OF THE UK'S LEGAL AID FUND.
LEGAL AID
WORKING TOWARD JUSTICE FOR ALL

THE FULL AMOUNT EVENTUALLY CAME TO £435,643, PLUS EXPENSES.

THIS MONEY WAS NEVER DECLARED TO THE LANCET AS IT SHOULD HAVE BEEN.

IT GETS WORSE.

NEARLY NINE MONTHS BEFORE THE PRESS CONFERENCE IN WHICH WAKEFIELD CALLED FOR SINGLE VACCINES...
MMR IS POISON
WE LOVE WAKEFIELD

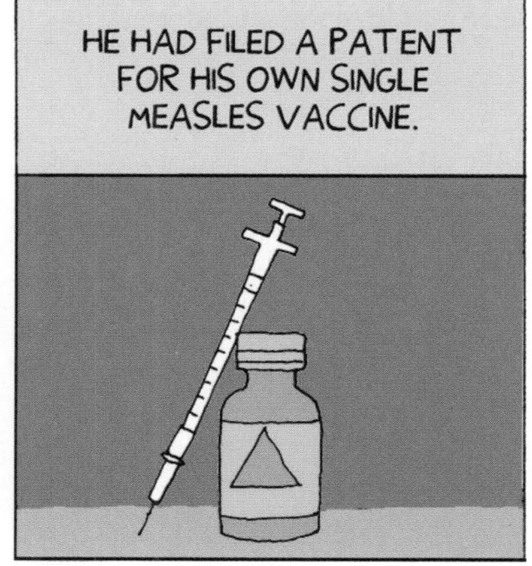
HE HAD FILED A PATENT FOR HIS OWN SINGLE MEASLES VACCINE.

THIS NEW VACCINE ONLY STOOD A CHANCE OF SUCCESS IF CONFIDENCE IN MMR WAS DAMAGED.

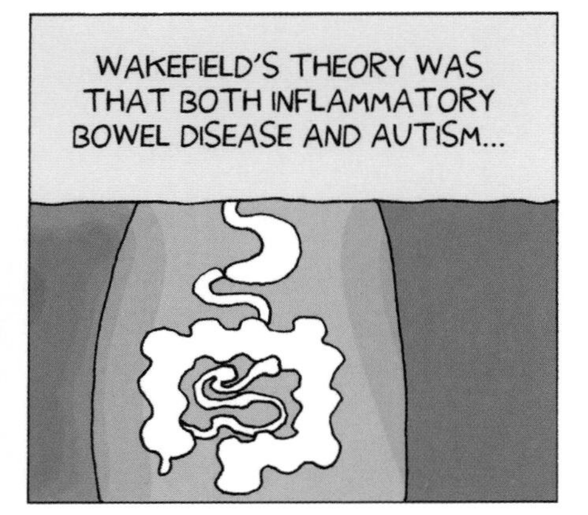

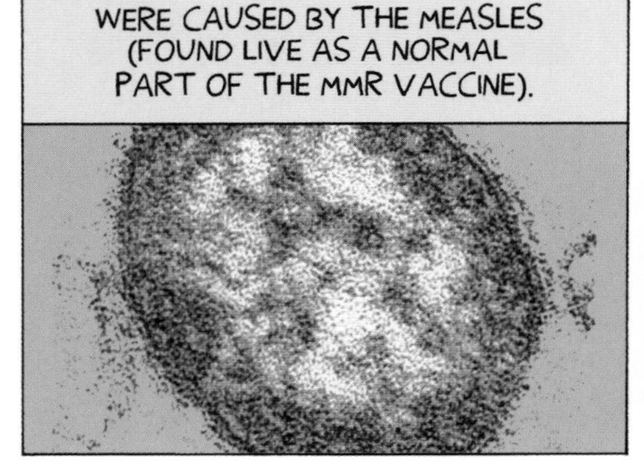

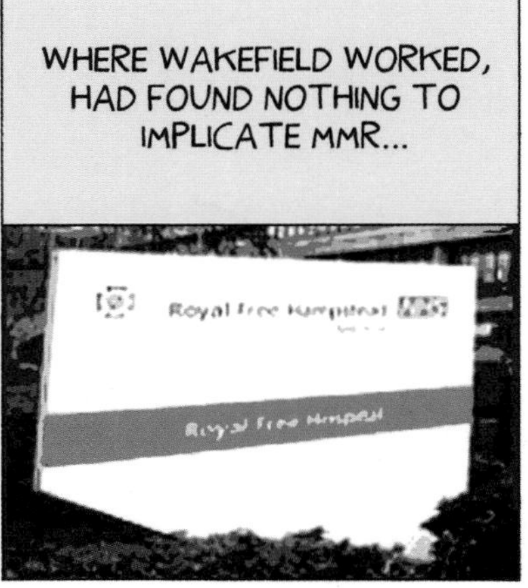

AND THAT A CLEAR MIS-MATCH EXISTED BETWEEN WAKEFIELD'S PUBLISHED PAPER...

AND THE NHS RECORDS OF THE CHILDREN IN QUESTION.

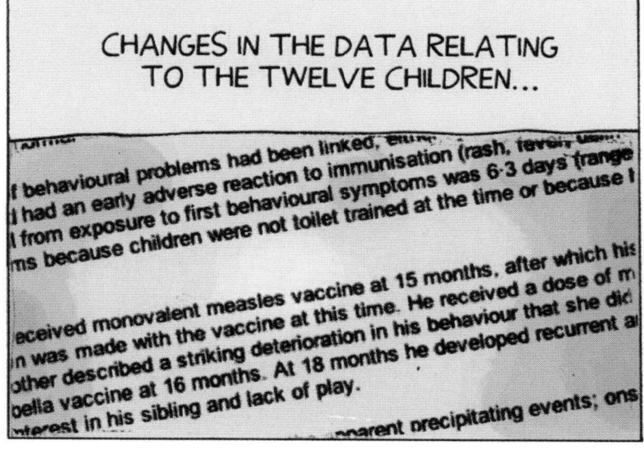
CHANGES IN THE DATA RELATING TO THE TWELVE CHILDREN...
f behavioural problems had been linked,
had an early adverse reaction to immunisation (rash,
from exposure to first behavioural symptoms was 6·3 days (range
ms because children were not toilet trained at the time or because
eceived monovalent measles vaccine at 15 months, after which his
n was made with the vaccine at this time. He received a dose of m
other described a striking deterioration in his behaviour that she did
bella vaccine at 16 months. At 18 months he developed recurrent a
nterest in his sibling and lack of play.
parent precipitating events; ons

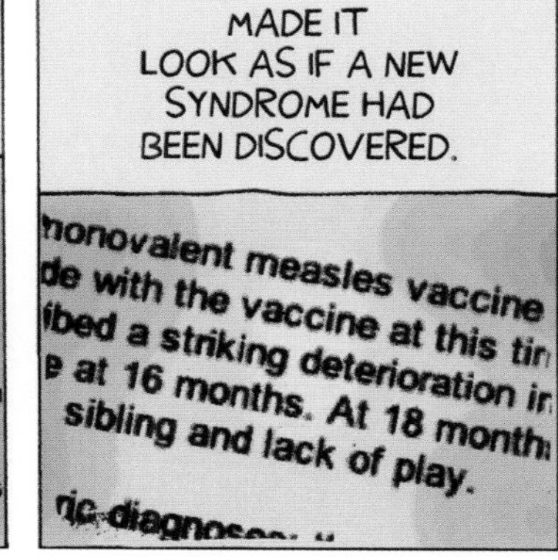
MADE IT LOOK AS IF A NEW SYNDROME HAD BEEN DISCOVERED.
monovalent measles vaccine
de with the vaccine at this tir
ibed a striking deterioration in
e at 16 months. At 18 month
sibling and lack of play.

THE GENERAL MEDICAL COUNCIL AND THE LANCET TOOK ACTION.
MMR
NEWS

THE LANCET RETRACTED THE PAPER.

THE GMC DISCIPLINARY PANEL,
WHICH SAT AND HEARD EVIDENCE
FOR 147 DAYS...

FOUND A LONG LIST
OF CHARGES AGAINST
WAKEFIELD PROVEN.

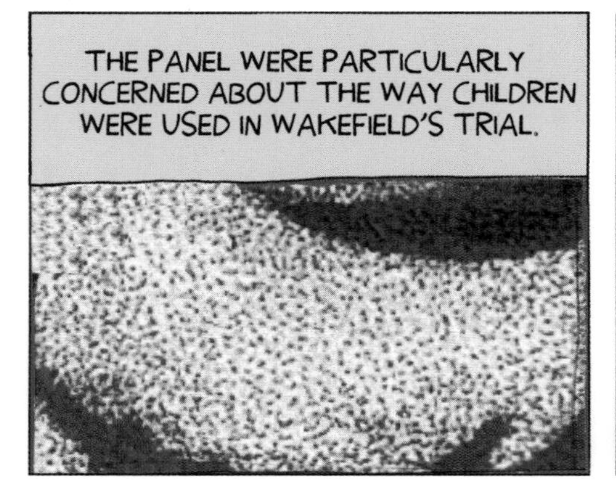
THE PANEL WERE PARTICULARLY
CONCERNED ABOUT THE WAY CHILDREN
WERE USED IN WAKEFIELD'S TRIAL.

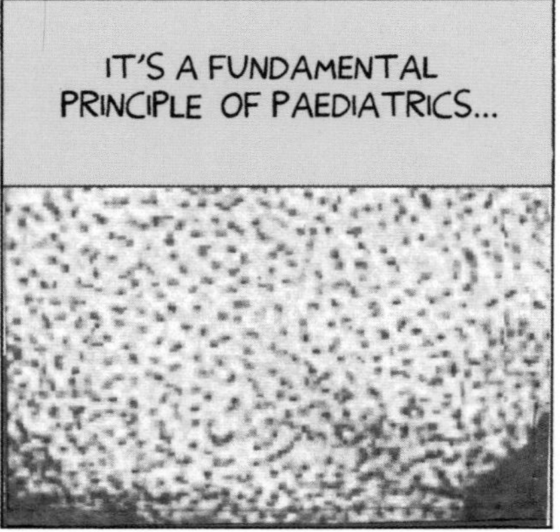
IT'S A FUNDAMENTAL
PRINCIPLE OF PAEDIATRICS...

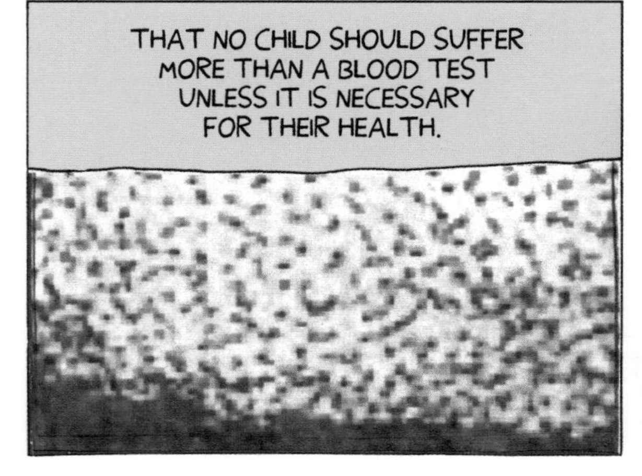
THAT NO CHILD SHOULD SUFFER
MORE THAN A BLOOD TEST
UNLESS IT IS NECESSARY
FOR THEIR HEALTH.

OBVIOUS, YOU
WOULD THINK.

HOWEVER, IN THE INTERESTS OF PROVING WAKEFIELD'S THEORY...

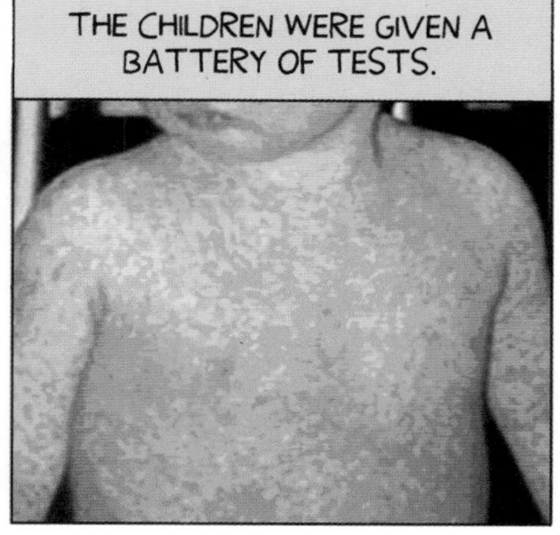
THE CHILDREN WERE GIVEN A BATTERY OF TESTS.

COLONOSCOPIES, LUMBAR PUNCTURES AND BARIUM MEALS. DISTRESSING PROCEDURES FOR ANY CHILD.

THE HOSPITAL'S ETHICS COMMITTEE WAS NOT PROPERLY CONSULTED.

ALL THIS EVIDENCE CRUSHED WAKEFIELD'S DEFENCE.

HE WAS FOUND BY THE GMC TO HAVE ACTED DISHONESTLY AND IRRESPONSIBLY.

WAKEFIELD NO LONGER WORKS AT THE ROYAL FREE HOSPITAL.

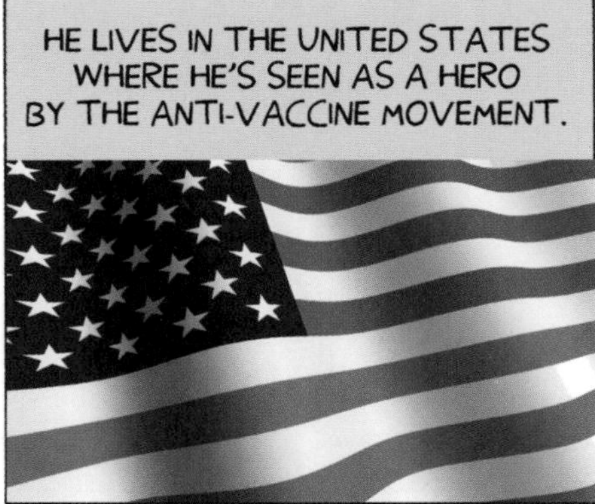
HE LIVES IN THE UNITED STATES WHERE HE'S SEEN AS A HERO BY THE ANTI-VACCINE MOVEMENT.

ENDORSED BY CELEBRITIES SUCH AS JIM CARREY...

AND JENNY MCCARTHY...

WHO ARE CONVINCED THAT A WORLDWIDE CONSPIRACY HAS BEEN ENGINEERED AGAINST WAKEFIELD...

BY VACCINE MANUFACTURERS AND THEIR PAID LACKEYS IN THE MEDIA AND SCIENTIFIC ESTABLISHMENT.
GOSH!
PLOTS AND SECRETS

FOR THE TEN YEARS THAT THE MMR CONTROVERSY RAGED...

THE MEDIA IN BRITAIN FELL OVER THEMSELVES TO PROMOTE AND SUPPORT DR. WAKEFIELD.
BLAH! BLAH!

MOST NEWSPAPERS, WITH ONLY A FEW EXCEPTIONS, UNCRITICALLY SWALLOWED WHOLE THE STUDY'S FEEBLE EVIDENCE...

SENSATIONALLY HIGHLIGHTING THE STORY...

WHILE DOWNPLAYING AND EVEN IGNORING STUDIES WHICH FOUND NO CONNECTION BETWEEN MMR AND AUTISM.

WAKEFIELD WAS FAR FROM ALONE IN CREATING THIS HEALTH SCARE.

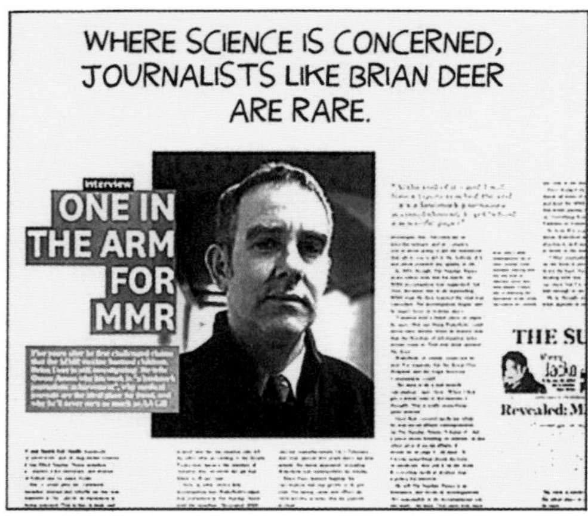
WHERE SCIENCE IS CONCERNED, JOURNALISTS LIKE BRIAN DEER ARE RARE.
interview
ONE IN THE ARM FOR MMR
THE SU
Revealed: M

INSTEAD WE HAVE POOR AND BIASED SCIENCE REPORTING FOR SENSATION-HUNGRY NEWSPAPERS...
DAILY SAME
BAN THIS SICK FILTH

WHO SPREAD ILL-INFORMED OPINIONS LIKE THEY WERE SOLID GOLD TRUTHS.
DAILY SAM

IN THE TWENTY-FIRST CENTURY, IS IT TOO MUCH TO ASK JOURNALISTS TO DO BASIC FACT-CHECKING?
DAILY SAM

AND THAT EDITORS WOULD ASSIGN SCIENCE STORIES TO REPORTERS WHO HAVE A KNOWLEDGE OF THE SUBJECT?
DAILY S

IS THAT TOO MUCH TO ASK?
END

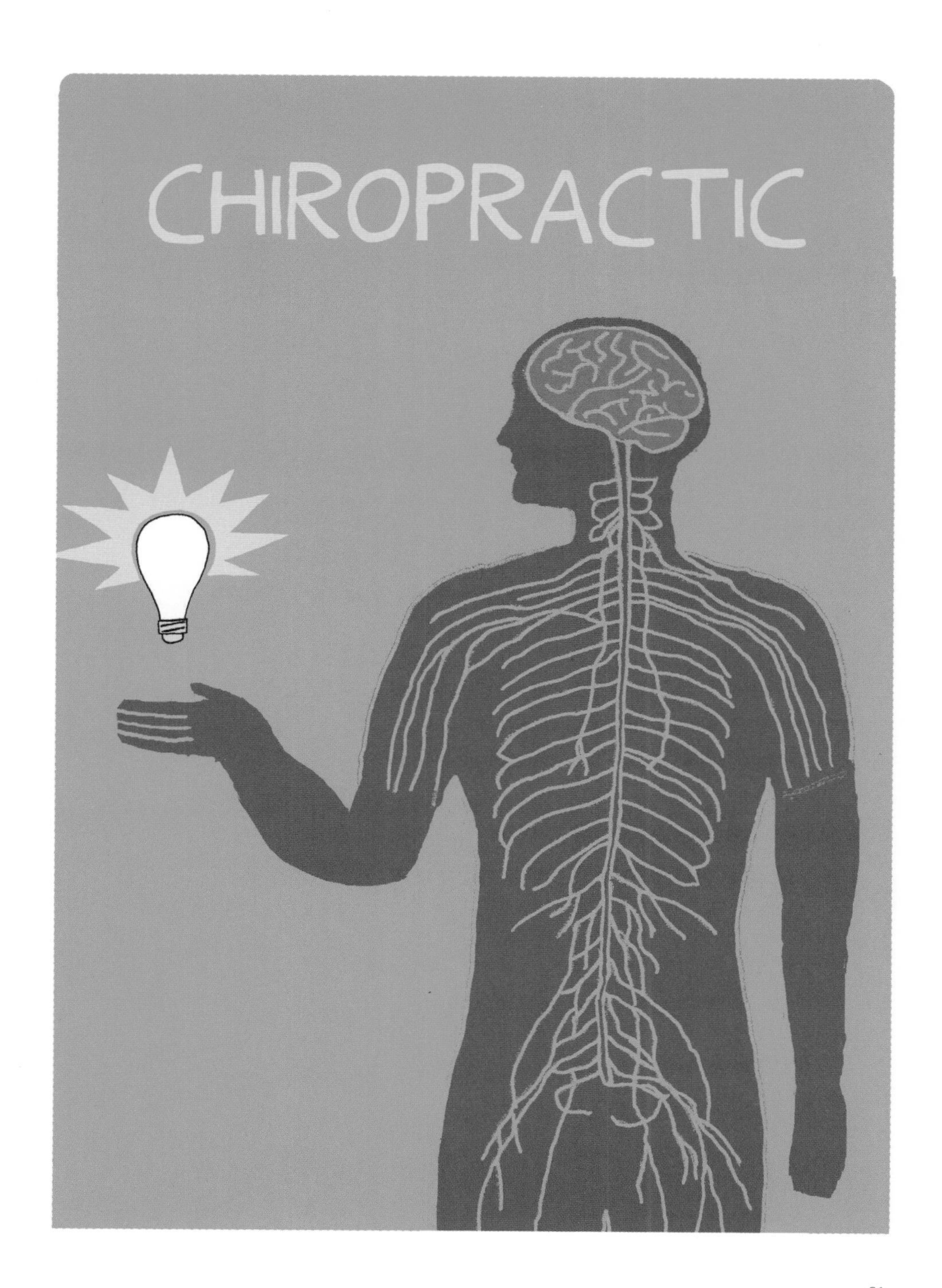
CHIROPRACTIC

OW!

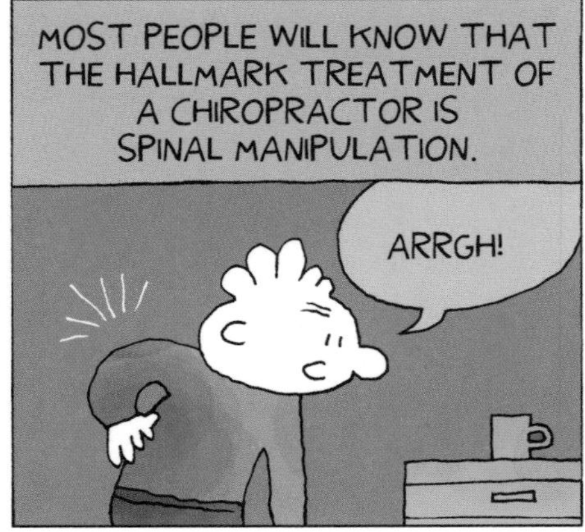
MOST PEOPLE WILL KNOW THAT THE HALLMARK TREATMENT OF A CHIROPRACTOR IS SPINAL MANIPULATION.
ARRGH!

VERY GOOD FOR THE BACK, YOU MIGHT THINK?

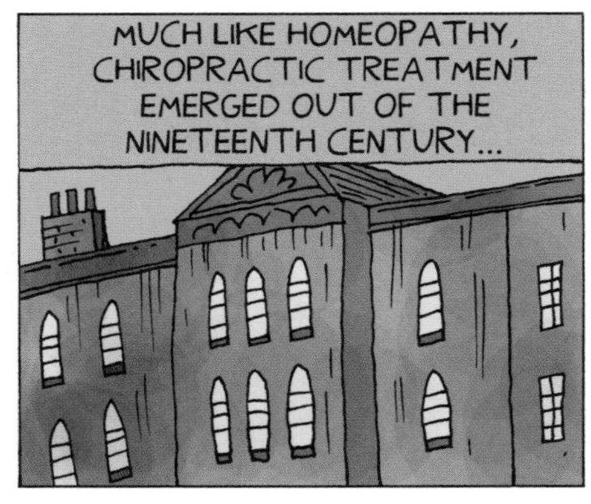
MUCH LIKE HOMEOPATHY, CHIROPRACTIC TREATMENT EMERGED OUT OF THE NINETEENTH CENTURY...

A TIME THAT WAS FULL OF ODDBALL IDEAS ABOUT HEALTH CARE.

THE TRADITION WAS FOUNDED BY DANIEL DAVID PALMER. BORN TORONTO, CANADA, 1845.

PALMER, WHO HAD NO FORMAL MEDICAL TRAINING...

WAS A DEVOTEE OF PHRENOLOGY, MAGNETIC THERAPY, AND OTHER PSEUDOSCIENTIFIC IDEAS OF THE ERA.

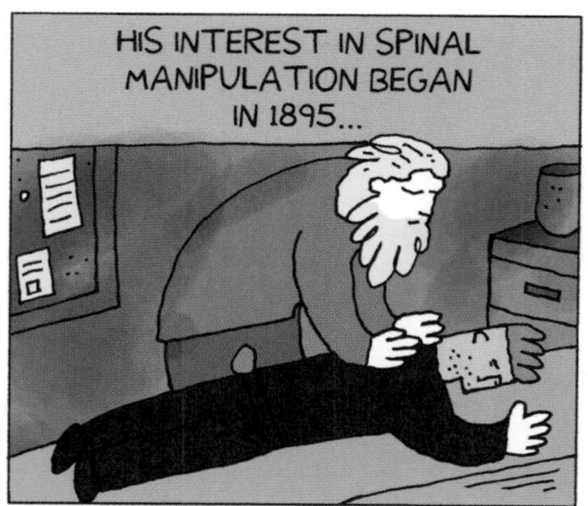
HIS INTEREST IN SPINAL MANIPULATION BEGAN IN 1895...

WHEN HE APPARENTLY CURED A JANITOR OF HIS DEAFNESS THROUGH MANIPULATION OF THE MAN'S VERTEBRAE.
HELLO!

PALMER THEN BECAME CONVINCED THAT SPINAL MANIPULATION COULD TREAT ALL MANNER OF HUMAN ILLS.
NO NEED TO SHOUT.

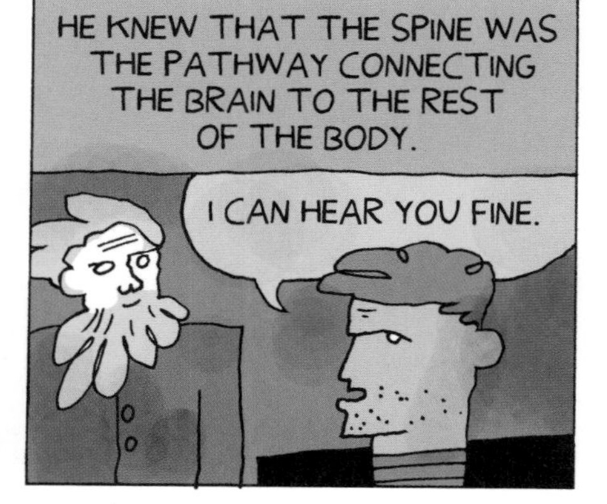
HE KNEW THAT THE SPINE WAS THE PATHWAY CONNECTING THE BRAIN TO THE REST OF THE BODY.
I CAN HEAR YOU FINE.

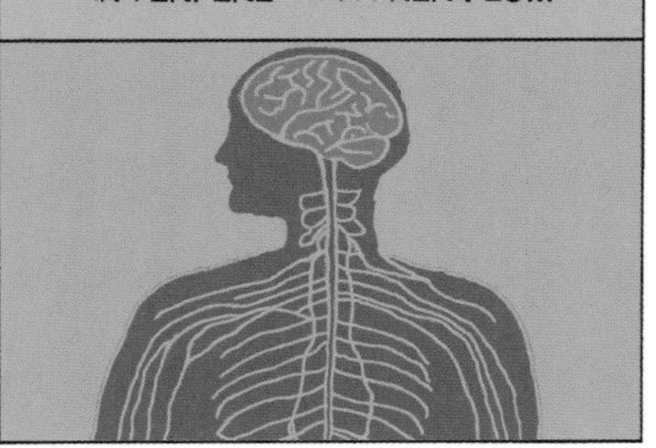
ACCORDING TO PALMER, DISPLACED VERTEBRAE WOULD INTERFERE WITH NERVES...

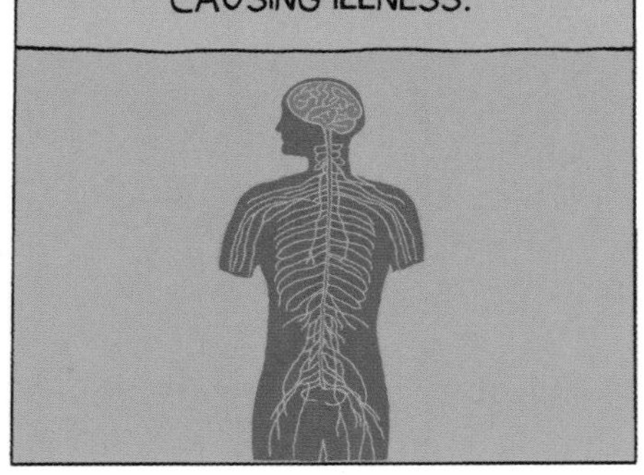
NEGATIVELY AFFECTING THE CONNECTING ORGANS, AND SO CAUSING ILLNESS.

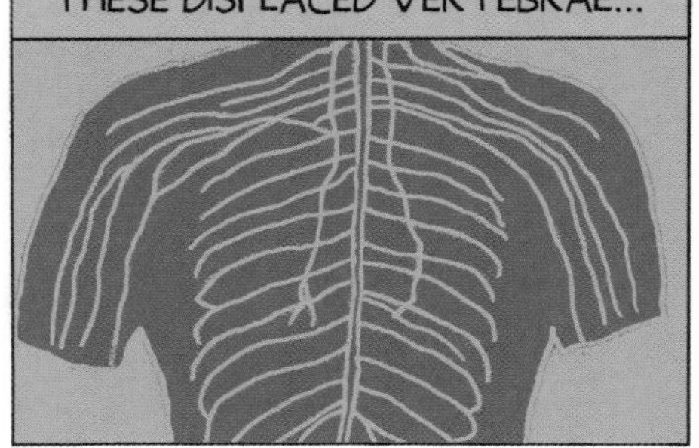
HIS BELIEF WAS THAT IF CHIROPRACTORS REALIGNED THESE DISPLACED VERTEBRAE...

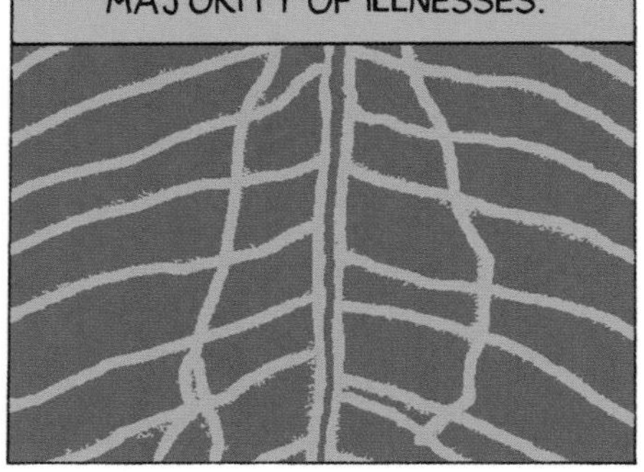
THEN THEY COULD CURE A MAJORITY OF ILLNESSES.

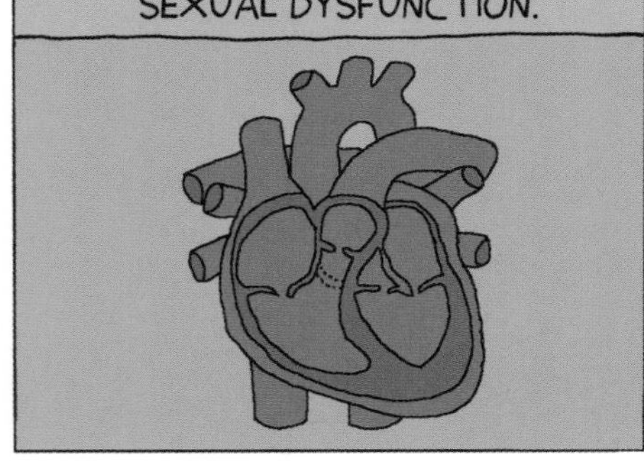
NOT JUST DEAFNESS, BUT EVERYTHING FROM HEART DISEASE, MEASLES, AND SEXUAL DYSFUNCTION.

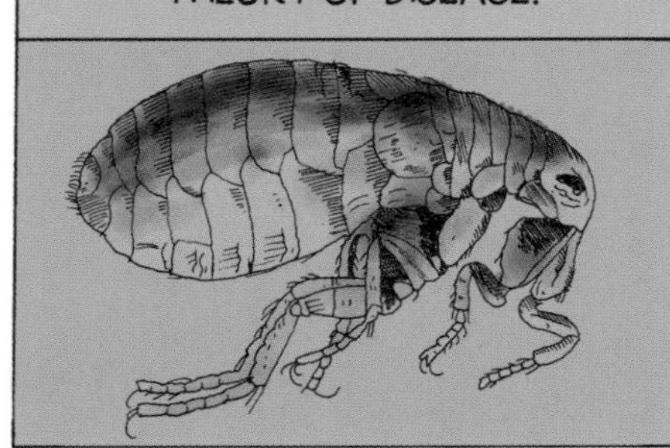
AN EXTRAORDINARY CLAIM, AND ONE THAT PAYS NO ATTENTION TO THE GERM THEORY OF DISEASE.

PALMER CALLED THESE SPINAL DISPLACEMENTS SUBLUXATIONS.

IN HIS THEORY, SUBLUXATIONS BLOCKED THE BODY'S INNATE INTELLIGENCE.

THE GUIDING ENERGY, CARRYING BOTH PHYSICAL AND PSYCHOLOGICAL SIGNIFICANCE.

NONE OF THIS IS AT ALL SCIENTIFIC AND IS BACKED BY NO MEDICAL EVIDENCE.

THERE WAS ALWAYS SOMETHING OF THE GURU ABOUT DAVID PALMER...

WHOSE MYSTICAL LEANINGS TINGED EARLY CHIROPRACTIC PRACTICE WITH RELIGION.

I AM THE FOUNDER OF CHIROPRACTIC, IN ITS SCIENCE, IN ITS ART, IN ITS PHILOSOPHY AND IN ITS RELIGION.

CHIROPRACTIC THERAPY HAD ITS FIRST MARTYR, WHEN IN 1906, PALMER WAS JAILED FOR PRACTISING MEDICINE WITHOUT A LICENCE...

AN EVENT THAT ONLY STRENGTHENED THE FAST-GROWING MOVEMENT.
BAH!

DAVID PALMER'S SON WAS BARTLETT JOSHUA PALMER.

B.J. PALMER TOOK OVER THE CHIROPRACTIC MOVEMENT AFTER HIS FATHER'S DEATH.
I'M IN CHARGE NOW.

THE ELDER PALMER DIED THREE WEEKS AFTER A STRANGE INCIDENT, WHEN...

HIS SON ACCIDENTALLY RAN OVER HIM IN A CAR.
ARRGH!

THE OFFICIAL CAUSE OF DEATH WAS TYPHOID, BUT BEING RUN OVER COULDN'T HAVE HELPED.

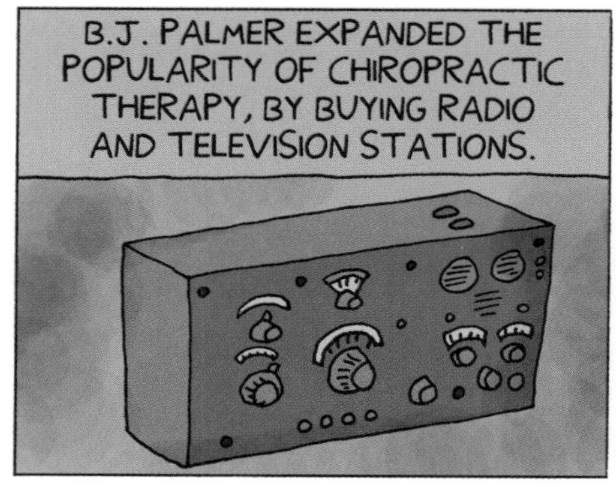
B.J. PALMER EXPANDED THE POPULARITY OF CHIROPRACTIC THERAPY, BY BUYING RADIO AND TELEVISION STATIONS.

IN 1924 HE INTRODUCED A MACHINE CALLED A NEUROCALOMETER.

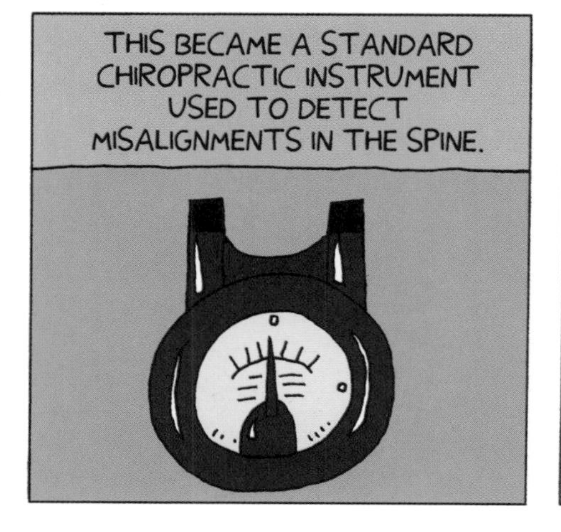
THIS BECAME A STANDARD CHIROPRACTIC INSTRUMENT USED TO DETECT MISALIGNMENTS IN THE SPINE.

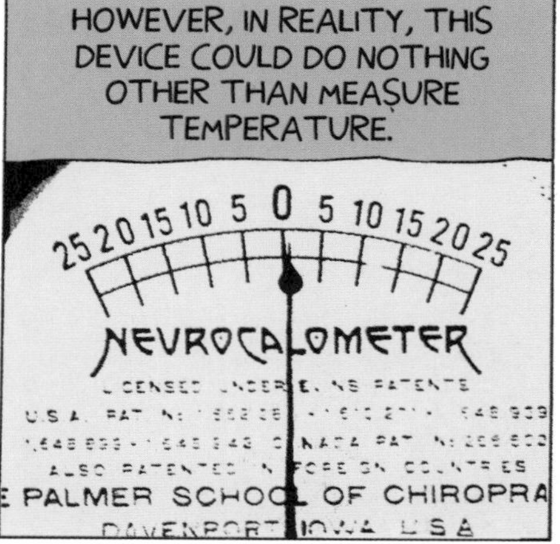
HOWEVER, IN REALITY, THIS DEVICE COULD DO NOTHING OTHER THAN MEASURE TEMPERATURE.
25 20 15 10 5 0 5 10 15 20 25
NEUROCALOMETER
PALMER SCHOOL OF CHIROPRA

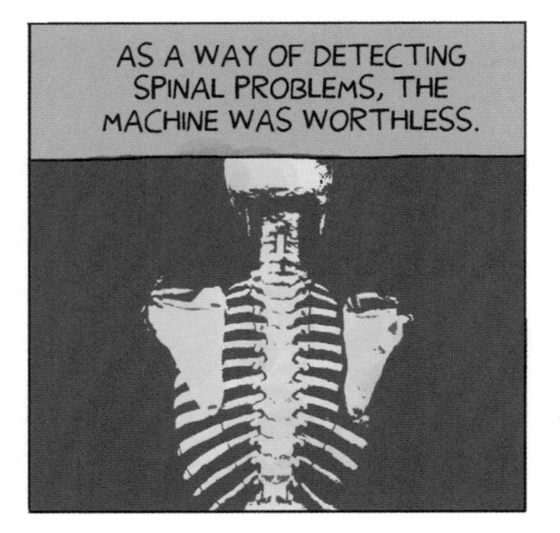
AS A WAY OF DETECTING SPINAL PROBLEMS, THE MACHINE WAS WORTHLESS.

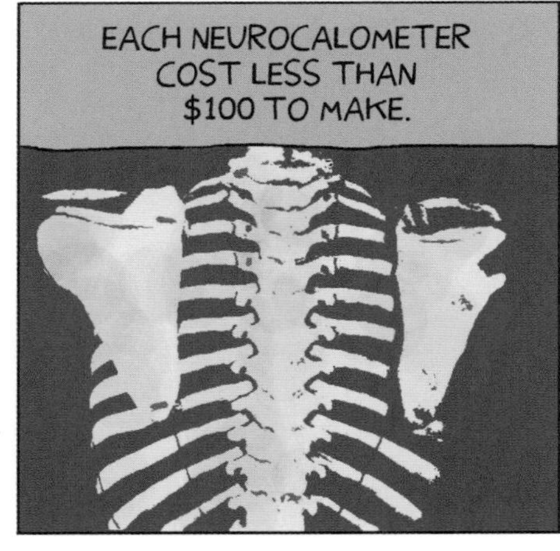
EACH NEUROCALOMETER COST LESS THAN $100 TO MAKE.

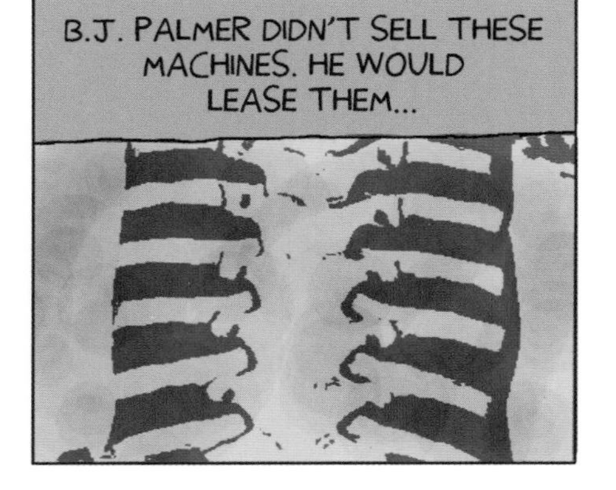
B.J. PALMER DIDN'T SELL THESE MACHINES. HE WOULD LEASE THEM...

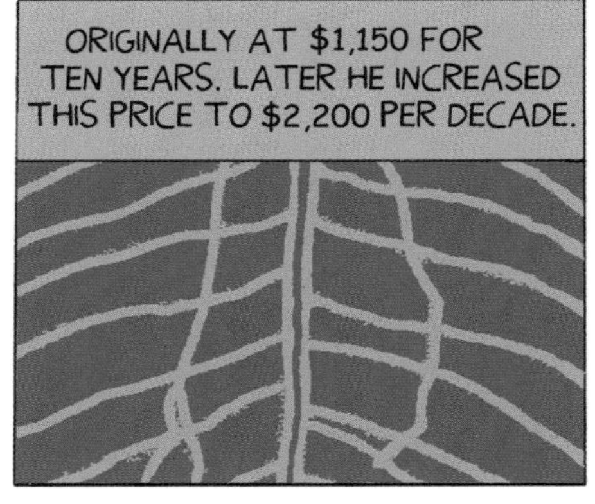
ORIGINALLY AT $1,150 FOR TEN YEARS. LATER HE INCREASED THIS PRICE TO $2,200 PER DECADE.

OVER 2,000 GRADUATES FROM HIS COLLEGE WERE OBLIGED TO LEASE A NEUROCALOMETER...
$
$
$

AT A COST THAT WOULD HAVE BOUGHT A HOUSE IN IOWA IN THE 1920s.

IN MODERN TIMES, CHIROPRACTIC HAS BECOME FRAUGHT WITH INTERNAL SCHISMS.

THERE IS A HUGE RANGE OF DIFFERENCE BETWEEN INDIVIDUAL CHIROPRACTORS.

THE MOVEMENT CAN BE DIVIDED INTO THREE GROUPS...

STRAIGHTS, MIXERS AND REFORMERS.

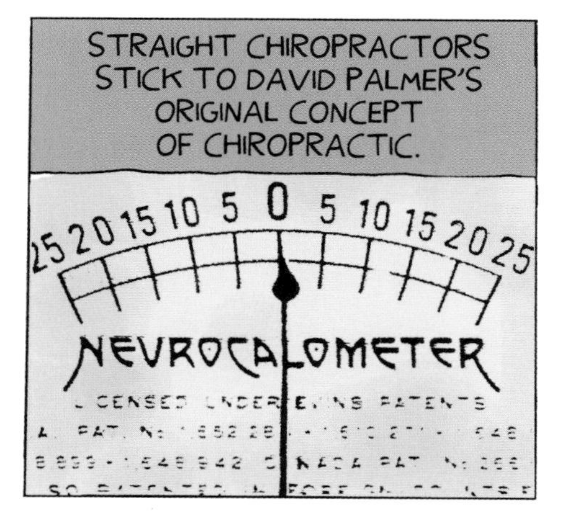
STRAIGHT CHIROPRACTORS STICK TO DAVID PALMER'S ORIGINAL CONCEPT OF CHIROPRACTIC.
25 20 15 10 5 0 5 10 15 20 25
NEVROCALOMETER

THEY ARE THE MOST EXTREME IN THEIR ANTI-SCIENCE VIEWS.

THEY ADVOCATE A PHILOSOPHICAL RATHER THAN A SCIENTIFIC BASIS FOR HEALTH CARE.

THEY REFER TO PHYSICIANS AS DRUG-PUSHERS AND DISPARAGE THE USE OF SURGERY.
TAKE THESE.
WOW!

THEY ARE CAREFUL NOT TO GIVE DISEASES NAMES, BUT CLAIM TO CURE THEM ANYWAY.

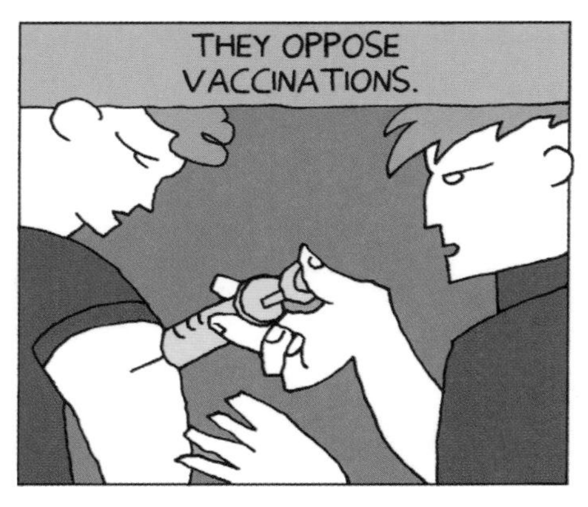
THEY OPPOSE VACCINATIONS.

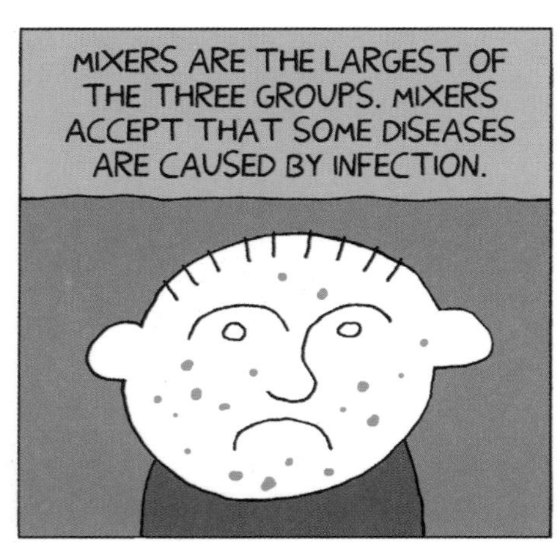
MIXERS ARE THE LARGEST OF THE THREE GROUPS. MIXERS ACCEPT THAT SOME DISEASES ARE CAUSED BY INFECTION.

THEY HAVE ADAPTED A MORE DEFINED JOB DESCRIPTION, AS BACK SPECIALISTS.
YOW!

THEY MAY AT FIRST GLANCE APPEAR MORE RATIONAL, BUT THE HEALTH CARE THEY PRACTISE ALONGSIDE CHIROPRACTIC...

TENDS TO BE UNPROVEN AND UNSCIENTIFIC. THEY USE METHODS SUCH AS...

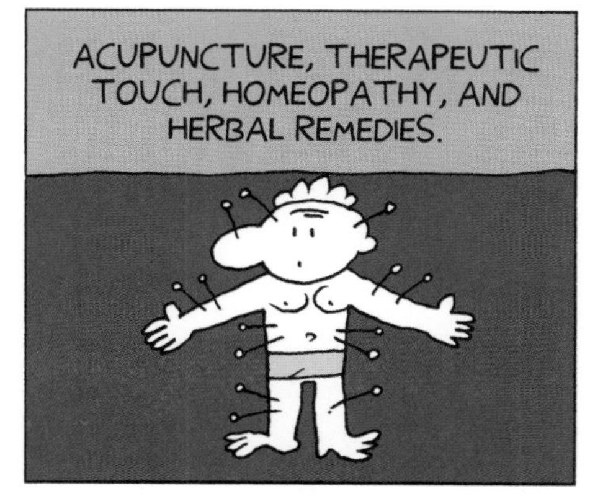
ACUPUNCTURE, THERAPEUTIC TOUCH, HOMEOPATHY, AND HERBAL REMEDIES.

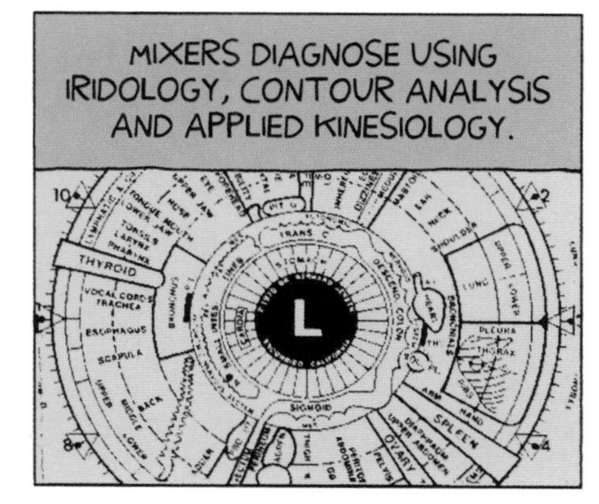
MIXERS DIAGNOSE USING IRIDOLOGY, CONTOUR ANALYSIS AND APPLIED KINESIOLOGY.
THYROID
L
SIGMOID
SPLEEN
OVARY

THIS APPROACH SHOWS AN ANTI-SCIENCE ATTITUDE AND A LACK OF SCIENTIFIC KNOWLEDGE.

THIS GROUP WOULD LIKE TO BE ACCEPTED BY MAINSTREAM MEDICINE.

MIXERS DON'T OPENLY OPPOSE IMMUNISATION, AS STRAIGHT CHIROPRACTORS DO.

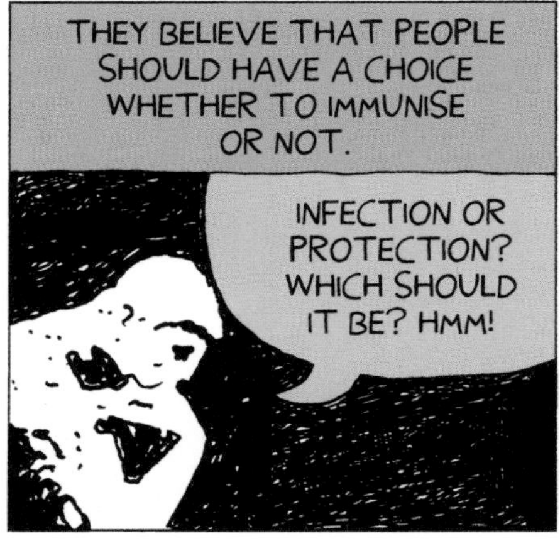
THEY BELIEVE THAT PEOPLE SHOULD HAVE A CHOICE WHETHER TO IMMUNISE OR NOT.
INFECTION OR PROTECTION? WHICH SHOULD IT BE? HMM!

THIS IGNORES THE FACT THAT IMMUNISATION IS INEFFECTIVE UNLESS THE MAJORITY OF THE POPULATION IS IMMUNISED.

MIXERS ASPIRE TO BE A GATEWAY PROFESSION FOR PATIENTS INTO HEALTH CARE.

EVEN THOUGH THEY LACK TRAINING IN GENERAL MEDICAL DIAGNOSIS.

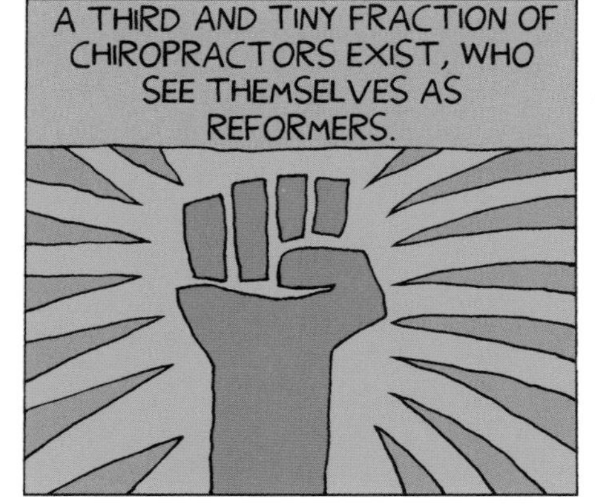
A THIRD AND TINY FRACTION OF CHIROPRACTORS EXIST, WHO SEE THEMSELVES AS REFORMERS.

THESE REFORMERS HAVE BEEN OPENLY CRITICAL OF THEIR FIELD.

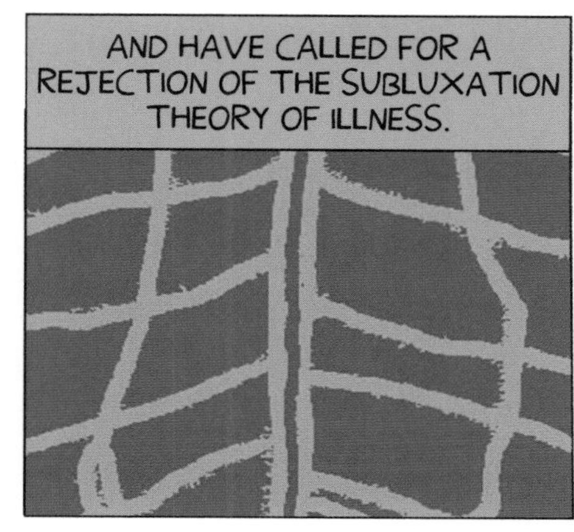
AND HAVE CALLED FOR A REJECTION OF THE SUBLUXATION THEORY OF ILLNESS.

THEY WISH TO END THE USE OF PSEUDOSCIENTIFIC PRACTICES IN THEIR DISCIPLINE...

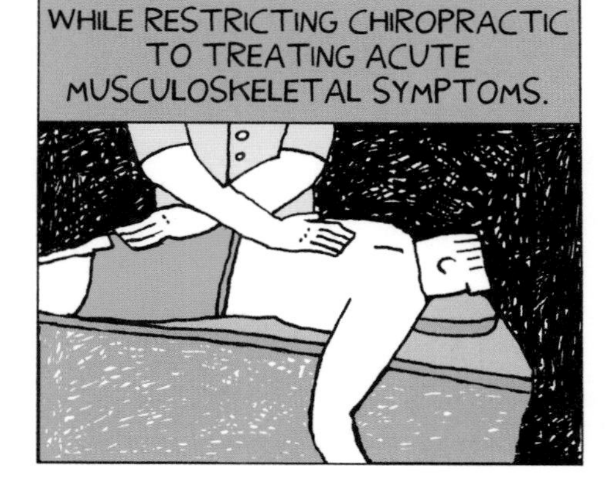
WHILE RESTRICTING CHIROPRACTIC TO TREATING ACUTE MUSCULOSKELETAL SYMPTOMS.

THESE REFORMS HAVE FAILED, BECAUSE CHIROPRACTIC APPEARS TO BE STILL ROOTED IN MYSTICISM...

AND HAS A REGULATORY INERTIA THAT HAS PROVED HARD TO CHANGE.

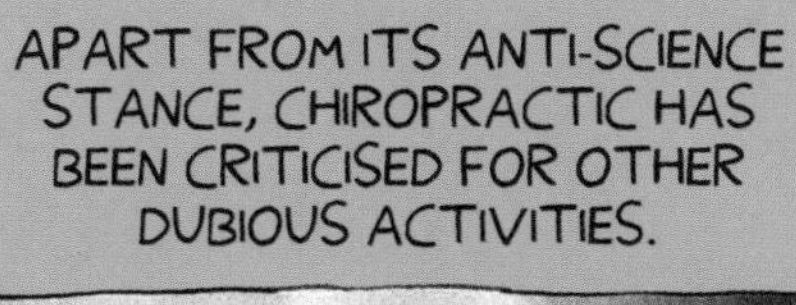
APART FROM ITS ANTI-SCIENCE STANCE, CHIROPRACTIC HAS BEEN CRITICISED FOR OTHER DUBIOUS ACTIVITIES.

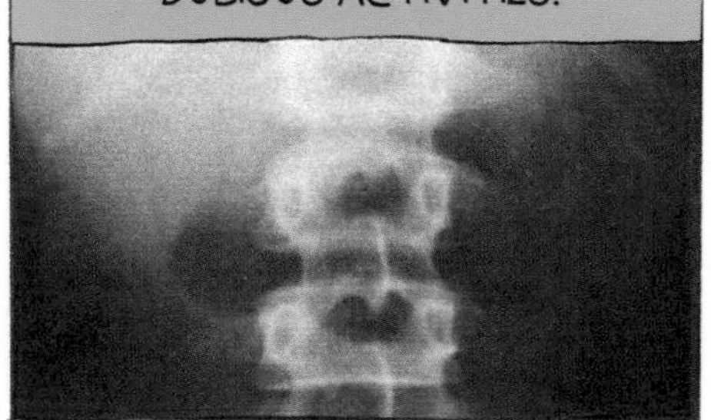

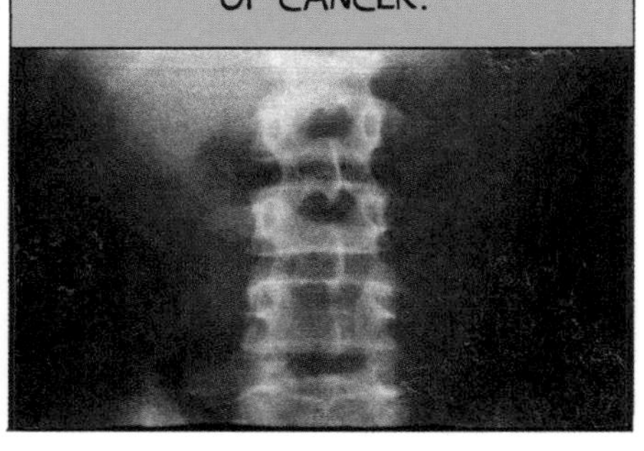
EVERY X-RAY TAKEN CARRIES WITH IT A MINUTE RISK OF CANCER.

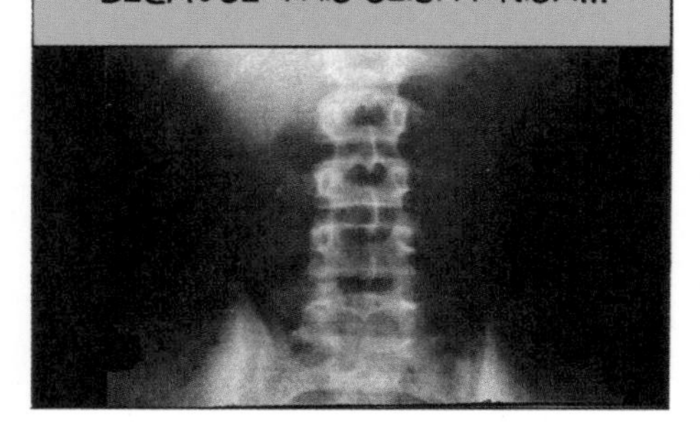
DESPITE THIS, CONVENTIONAL DOCTORS STILL USE X-RAYS, BECAUSE THIS SLIGHT RISK...

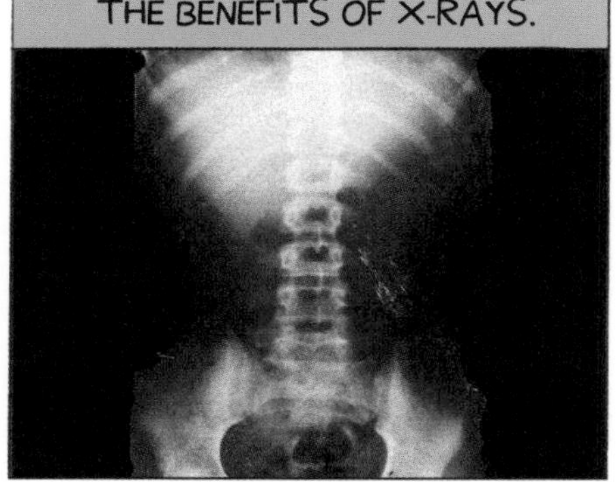
IS FAR OUTWEIGHED BY THE BENEFITS OF X-RAYS.

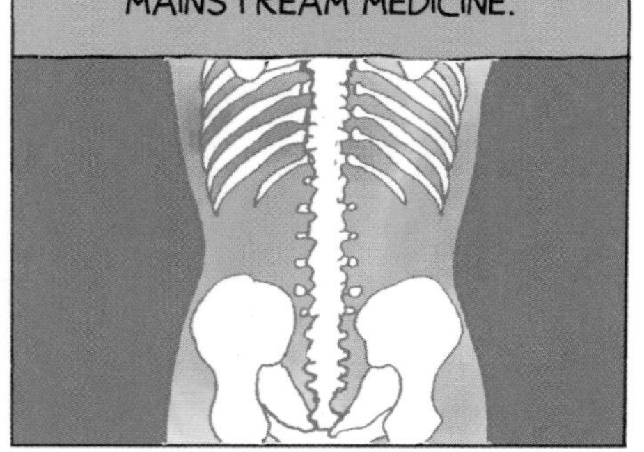
EVEN SO, EXPOSURE TO X-RAYS IS KEPT TO A MINIMUM BY MAINSTREAM MEDICINE.

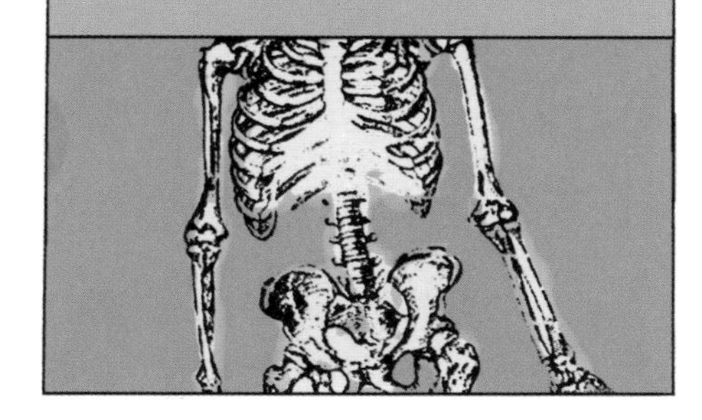
BY CONTRAST, A CHIROPRACTOR MAY X-RAY THE SAME PATIENT SEVERAL TIMES A YEAR.

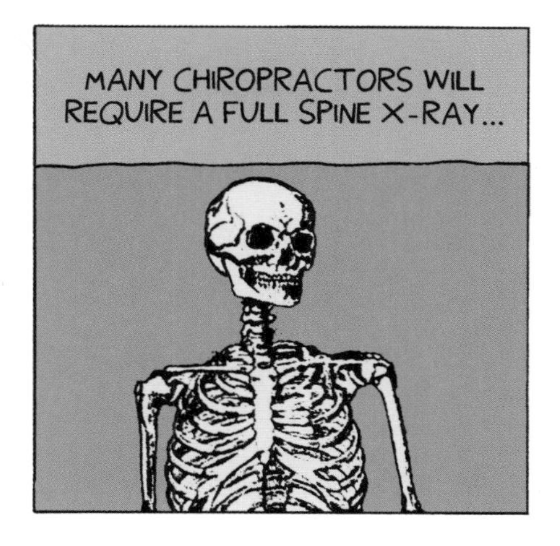
MANY CHIROPRACTORS WILL REQUIRE A FULL SPINE X-RAY...

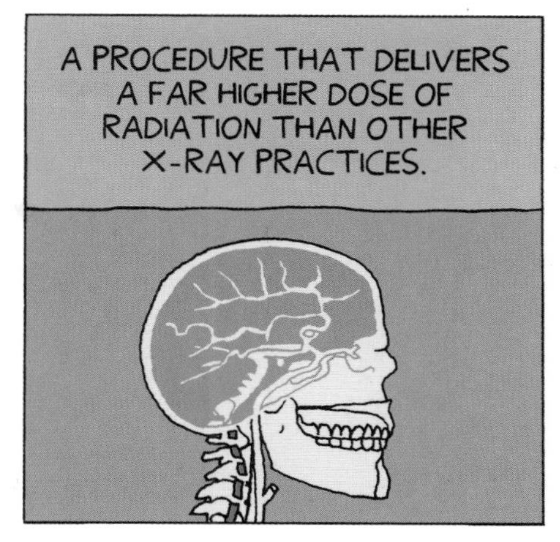
A PROCEDURE THAT DELIVERS A FAR HIGHER DOSE OF RADIATION THAN OTHER X-RAY PRACTICES.

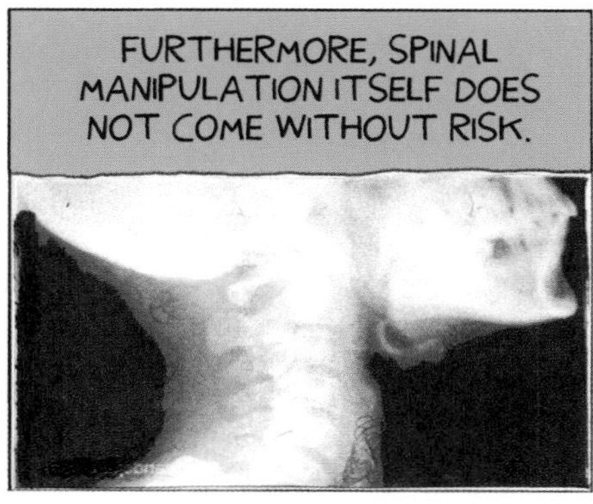
FURTHERMORE, SPINAL MANIPULATION ITSELF DOES NOT COME WITHOUT RISK.

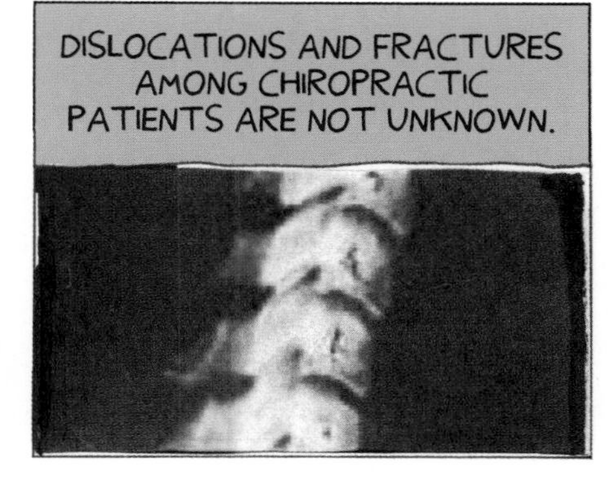
DISLOCATIONS AND FRACTURES AMONG CHIROPRACTIC PATIENTS ARE NOT UNKNOWN.

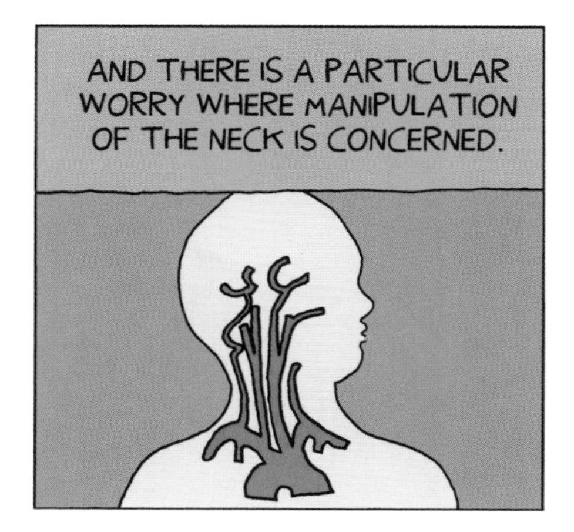
AND THERE IS A PARTICULAR WORRY WHERE MANIPULATION OF THE NECK IS CONCERNED.

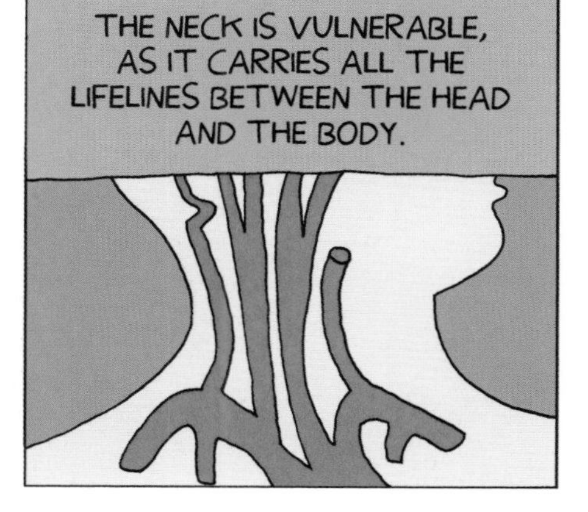
THE NECK IS VULNERABLE, AS IT CARRIES ALL THE LIFELINES BETWEEN THE HEAD AND THE BODY.

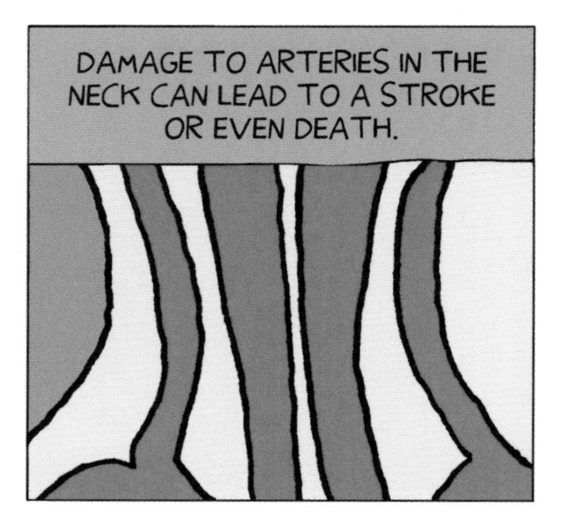
DAMAGE TO ARTERIES IN THE NECK CAN LEAD TO A STROKE OR EVEN DEATH.

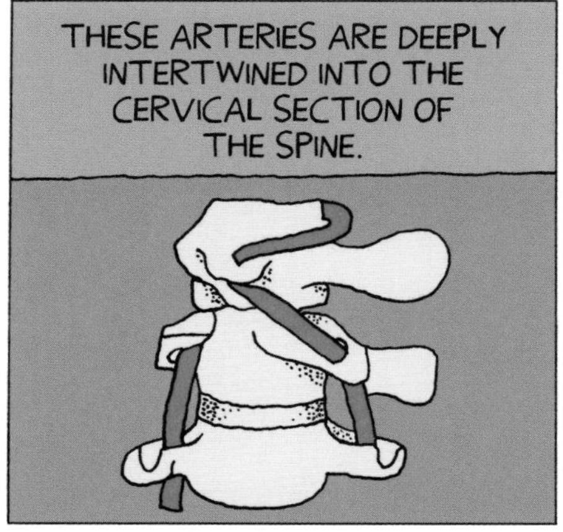
THESE ARTERIES ARE DEEPLY INTERTWINED INTO THE CERVICAL SECTION OF THE SPINE.

BECAUSE THERE'S OFTEN A DELAY BETWEEN THE INITIAL DAMAGE AND THE RESULTING STROKE...

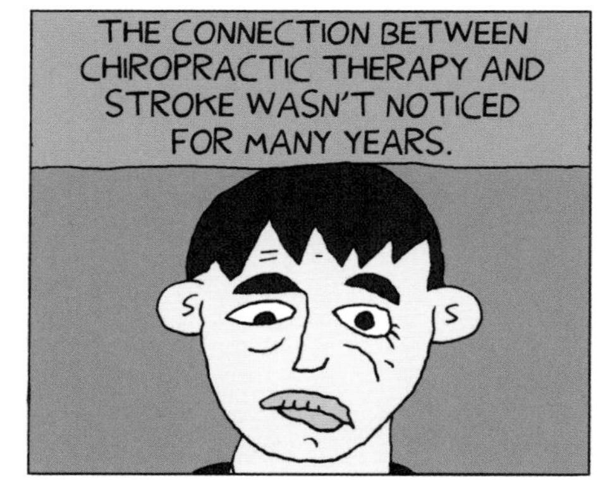
THE CONNECTION BETWEEN CHIROPRACTIC THERAPY AND STROKE WASN'T NOTICED FOR MANY YEARS.

IN RECENT TIMES, HOWEVER, MANY EXAMPLES HAVE COME TO LIGHT.

KRISTI BEDENBAUGH, 24, OF LITTLE MOUNTAIN, SOUTH CAROLINA, USA.

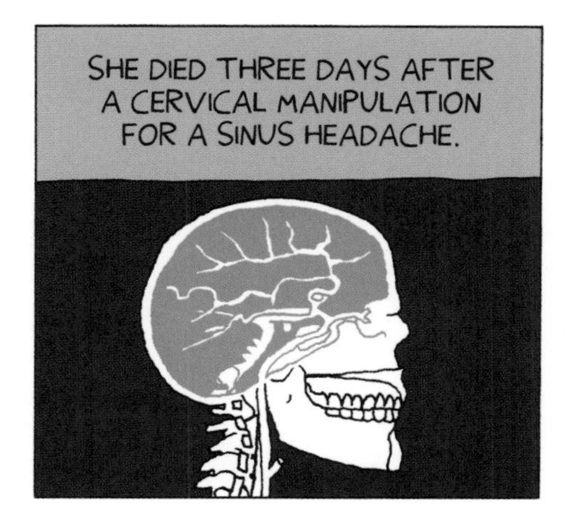
SHE DIED THREE DAYS AFTER A CERVICAL MANIPULATION FOR A SINUS HEADACHE.

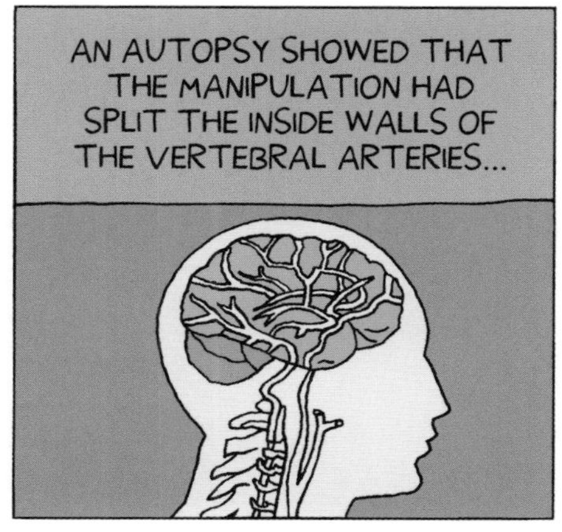
AN AUTOPSY SHOWED THAT THE MANIPULATION HAD SPLIT THE INSIDE WALLS OF THE VERTEBRAL ARTERIES...

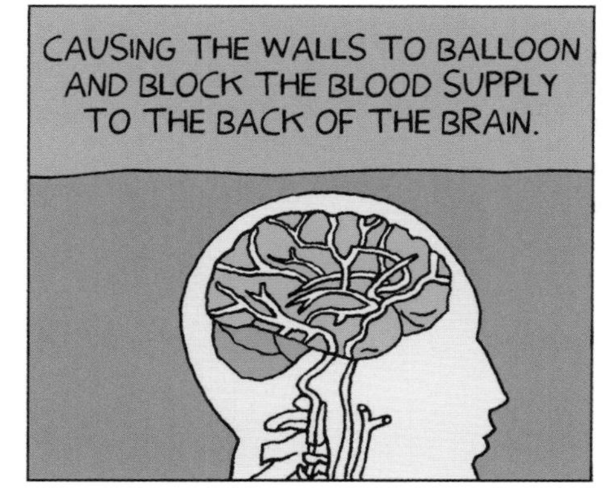
CAUSING THE WALLS TO BALLOON AND BLOCK THE BLOOD SUPPLY TO THE BACK OF THE BRAIN.

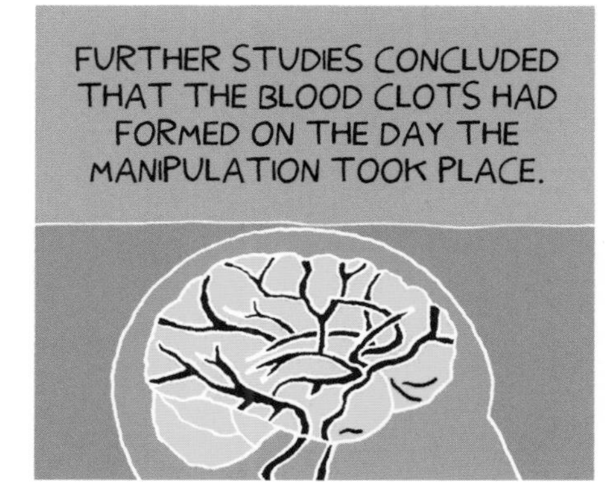
FURTHER STUDIES CONCLUDED THAT THE BLOOD CLOTS HAD FORMED ON THE DAY THE MANIPULATION TOOK PLACE.

LAURIE JEAN MATHIASON OF SASKATOON, SASKATCHEWAN, CANADA, DIED IN 1998, AGE 20...

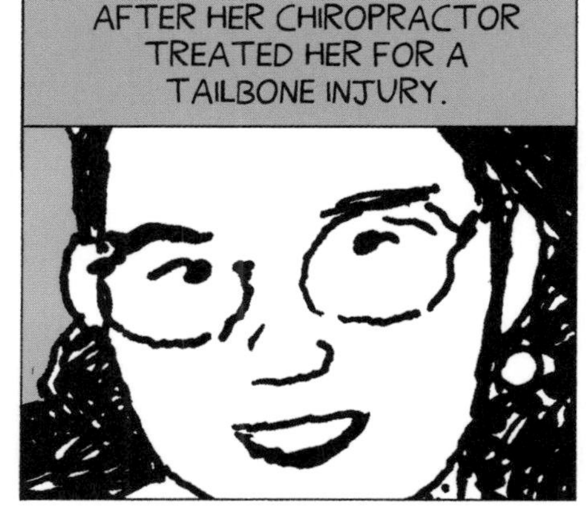
AFTER HER CHIROPRACTOR TREATED HER FOR A TAILBONE INJURY.

LAURIE'S MOTHER, SHARON MATHIASON.

THE TWIST WAS SO VIOLENT THAT IT TORE HER ARTERY CLEAN THROUGH.

AT THE HOSPITAL WE WERE BOMBARDED WITH DOCTORS COMING INTO THE WAITING ROOM, SAYING...

DON'T YOU KNOW THAT IF YOU GO TO A CHIROPRACTOR, DON'T LET THEM TOUCH YOU ABOVE THE NECK.

BRITTMARIE HARWE, 40, OF WETHERSFIELD, CONNECTICUT, USA.

SHE RECEIVED AN OUT-OF-COURT SETTLEMENT OF $900,000 AFTER A 1993 MANIPULATION PARALYSED ONE OF HER VOCAL CORDS.

THESE STORIES ARE ALARMING. BUT HOW LARGE IS THE RISK FROM CHIROPRACTIC NECK MANIPULATION?

THE FEW STUDIES THAT HAVE BEEN DONE SUGGEST THAT THE RISK IS SMALL, BUT REAL.

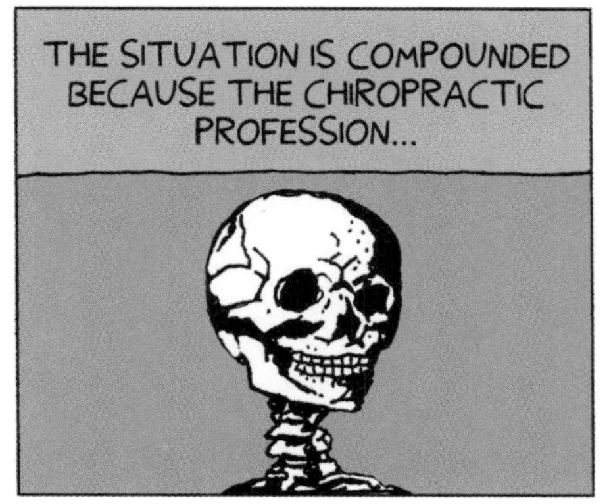
THE SITUATION IS COMPOUNDED BECAUSE THE CHIROPRACTIC PROFESSION...

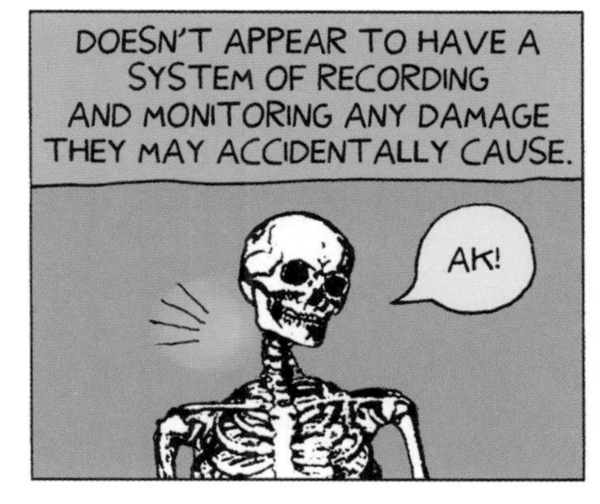
DOESN'T APPEAR TO HAVE A SYSTEM OF RECORDING AND MONITORING ANY DAMAGE THEY MAY ACCIDENTALLY CAUSE.
AK!

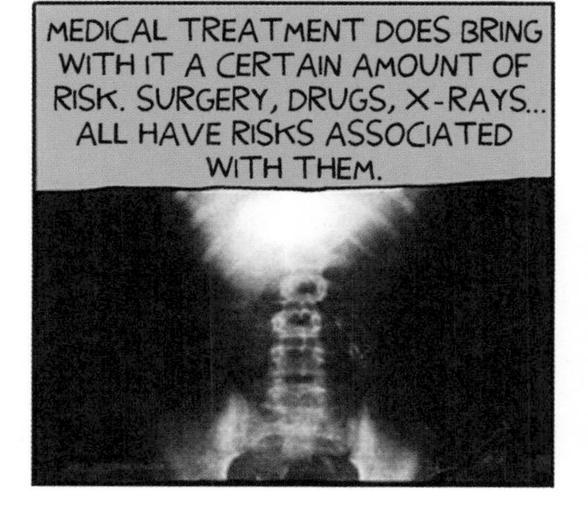
MEDICAL TREATMENT DOES BRING WITH IT A CERTAIN AMOUNT OF RISK. SURGERY, DRUGS, X-RAYS... ALL HAVE RISKS ASSOCIATED WITH THEM.

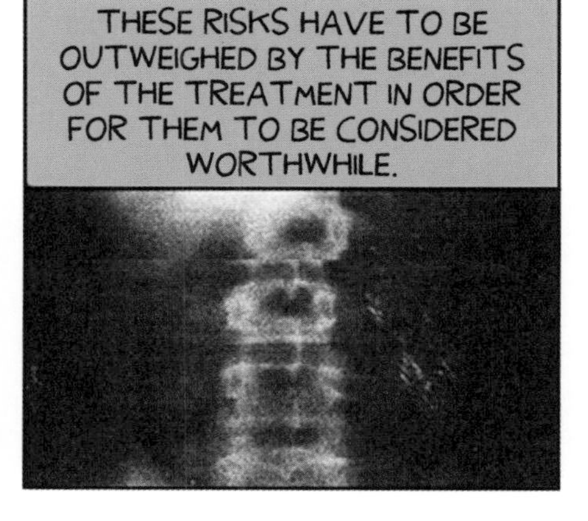
THESE RISKS HAVE TO BE OUTWEIGHED BY THE BENEFITS OF THE TREATMENT IN ORDER FOR THEM TO BE CONSIDERED WORTHWHILE.

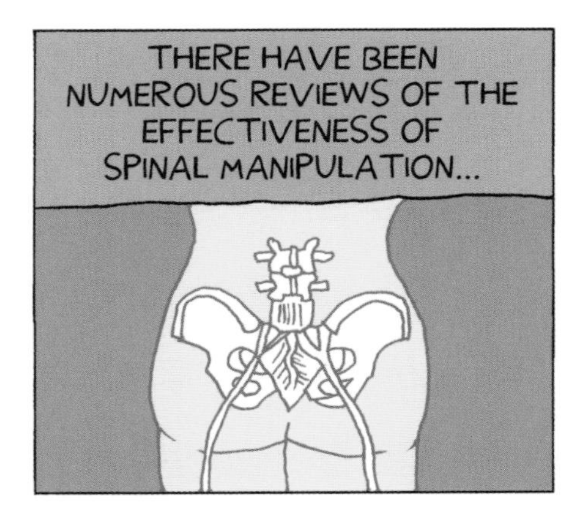
THERE HAVE BEEN NUMEROUS REVIEWS OF THE EFFECTIVENESS OF SPINAL MANIPULATION...

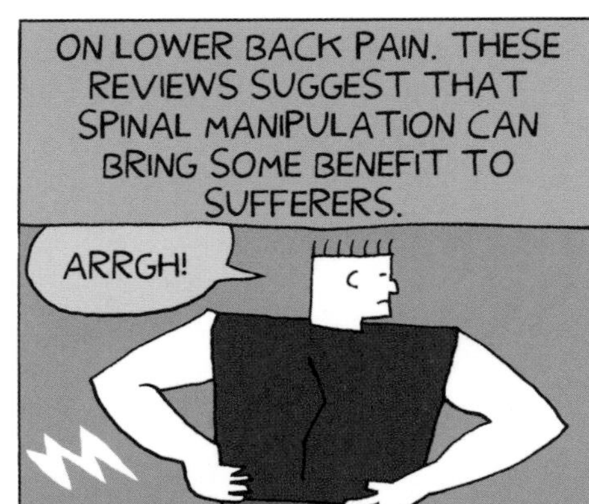
ON LOWER BACK PAIN. THESE REVIEWS SUGGEST THAT SPINAL MANIPULATION CAN BRING SOME BENEFIT TO SUFFERERS.
ARRGH!

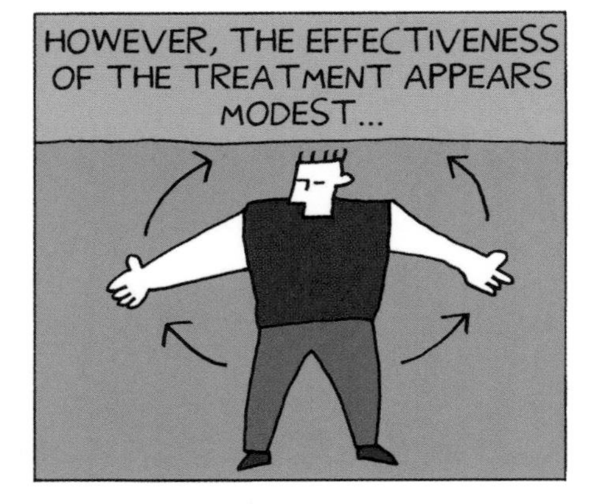
HOWEVER, THE EFFECTIVENESS OF THE TREATMENT APPEARS MODEST...

AND IS CERTAINLY NO BETTER THAN TREATMENT GIVEN BY CONVENTIONAL MEDICINE...

WHERE DOCTORS MAY RECOMMEND PHYSIOTHERAPY, EXERCISE, OR PRESCRIBE ANTI-INFLAMMATORY DRUGS.

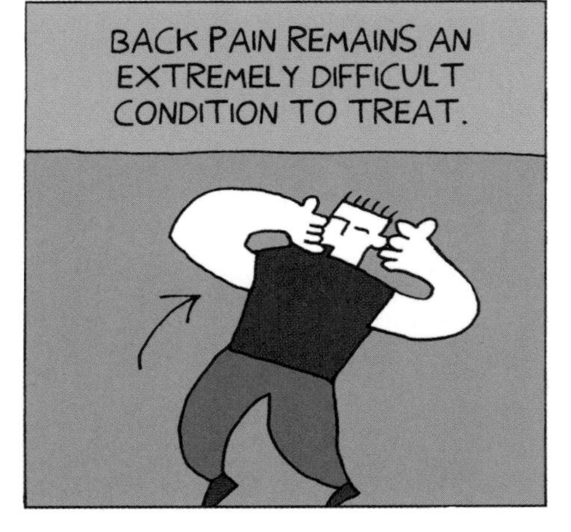
BACK PAIN REMAINS AN EXTREMELY DIFFICULT CONDITION TO TREAT.

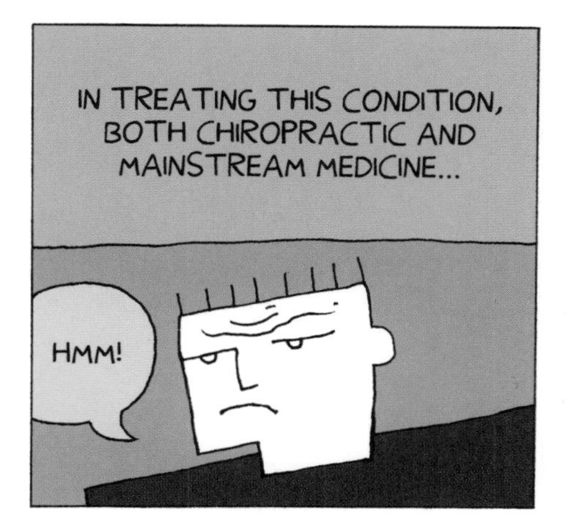
IN TREATING THIS CONDITION, BOTH CHIROPRACTIC AND MAINSTREAM MEDICINE...
HMM!

APPEAR TO BE AS EFFECTIVE, OR AS NON-EFFECTIVE, AS EACH OTHER.
STILL HURTS.

HOWEVER, CONVENTIONAL TREATMENT HAS SEVERAL ADVANTAGES OVER CHIROPRACTIC THERAPY.

CHIROPRACTIC THERAPY CAN BE EXPENSIVE.
SO HOW MUCH IS IT, THEN?

BILLS WILL INEVITABLY MOUNT UP OVER REPEATED SESSIONS.
EIGHT HUNDRED POUNDS, PLEASE.
GAK!

BY COMPARISON, EXERCISE IS FREE AND IBUPROFEN CHEAP.

OTHER PROFESSIONS EXIST THAT OFFER EQUIVALENT SERVICES...

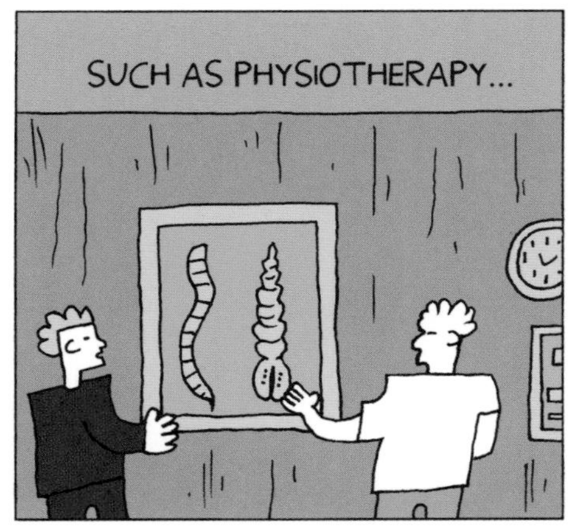
SUCH AS PHYSIOTHERAPY...

WHERE YOU WON'T BE GIVEN UNSCIENTIFIC ADVICE.
VACCINATIONS ARE QUITE UNNECESSARY FOR CHILDREN.

WHILE I'VE BEEN DRAWING THIS CHAPTER, I'VE HEARD FROM A FEW PEOPLE WHO...

HAVE TOLD ME THAT CHIROPRACTIC THERAPY HAS EASED OR EVEN CURED THEIR BACK PAIN.

TO WHICH I WOULD SAY THAT THEIR SUBJECTIVE EXPERIENCE, HOWEVER POSITIVE...

DOES NOT TRUMP THE WHOLE OF SCIENCE.

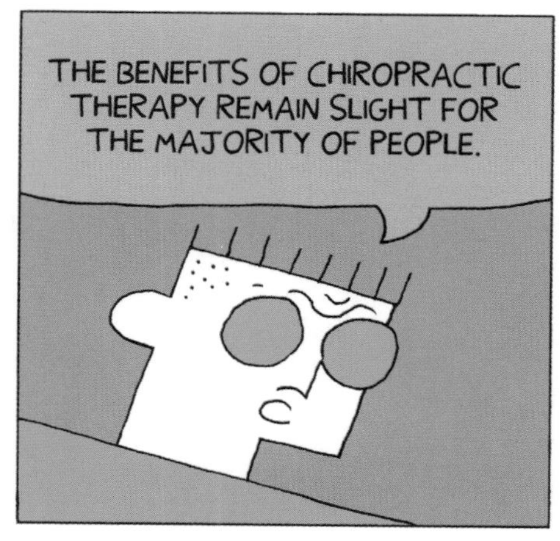
THE BENEFITS OF CHIROPRACTIC THERAPY REMAIN SLIGHT FOR THE MAJORITY OF PEOPLE.

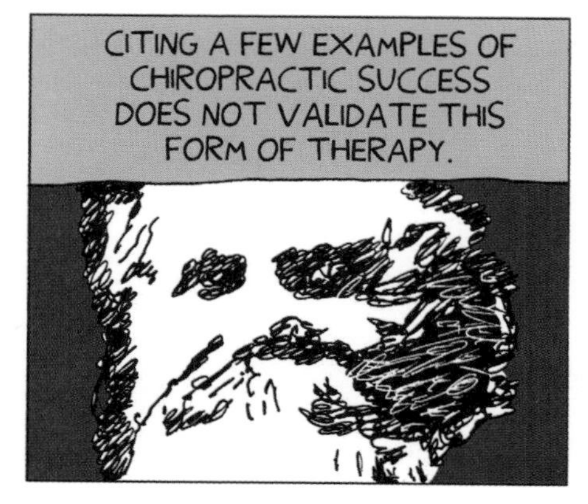
CITING A FEW EXAMPLES OF CHIROPRACTIC SUCCESS DOES NOT VALIDATE THIS FORM OF THERAPY.

A REALLY EFFECTIVE CURE FOR BACK PAIN HAS YET TO BE FOUND.

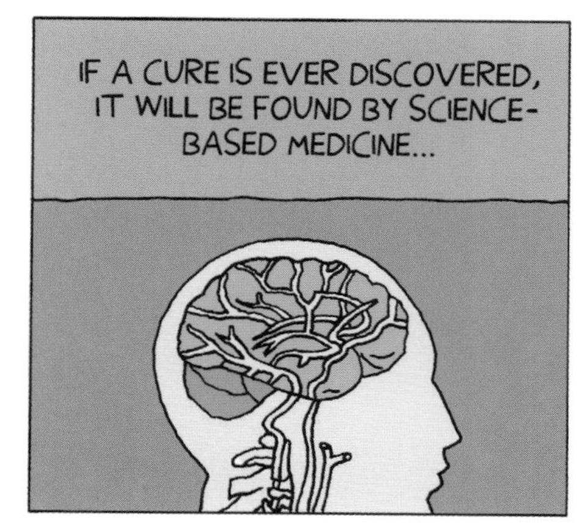
IF A CURE IS EVER DISCOVERED, IT WILL BE FOUND BY SCIENCE-BASED MEDICINE...

AND NOT FROM AN UNSCIENTIFIC AND MYSTICAL DISCIPLINE LIKE CHIROPRACTIC THERAPY.
END

THE
MOON
HOAX

THE APOLLO ELEVEN SPACE FLIGHT LANDED THE FIRST HUMANS ON EARTH'S MOON, JULY 20, 1969.

THE MISSION, CARRIED OUT BY THE UNITED STATES, IS CONSIDERED A MAJOR ACCOMPLISHMENT IN HUMAN EXPLORATION...

AND AS A VICTORY IN THE SPACE RACE WITH THE SOVIET UNION.

HOWEVER, IN THE DECADES SINCE, A VOCAL MINORITY HAS APPEARED WHICH CLAIMS THAT THE MOON LANDINGS...

NEVER HAPPENED.

THEY MAKE THE CLAIM THAT THE MOON LANDINGS WERE SHOT ON A FILM SET.

THE EVIDENCE THEY PRESENT DOESN'T AMOUNT TO MUCH AND IS EASILY REFUTED.
THE DENVER POST
Americans Walk on Moon; Lunar Takeoff Successful
'One Small Step for Man . . .'
World Sees Astronauts Plant Flag
'They Came in Peace . . .'
'. . . A Giant Leap for Mankind'

BUT THE MERE FACT THAT THESE ACCUSATIONS KEEP BEING MADE HAS CREATED DOUBT WHERE PREVIOUSLY THERE WAS NONE.

WHY ARE THERE NO STARS IN THE MOON PHOTOS?

THE MOON'S SURFACE PRESENTS A LOT OF GLARE FOR A PHOTOGRAPHER.

THERE'S NO ATMOSPHERE TO DIFFUSE THE SUNLIGHT AND THIS CREATES A LANDSCAPE OF HARSH CONTRASTS.

IT'S HARD TO CAPTURE SOMETHING VERY BRIGHT...

AND SOMETHING VERY DIM, ON THE SAME PIECE OF FILM.

ASTRONAUTS IN THEIR SUNLIT SPACE SUITS WALKING ACROSS THE BRIGHT LUNAR LANDSCAPE.

CAMERA SETTINGS FOR THIS IMAGE WOULD RENDER BACKGROUND STARS TOO FAINT TO BE SEEN.

IF THE SUN IS THE ONLY LIGHT SOURCE, HOW IS IT THAT THE ASTRONAUTS CAN STILL BE SEEN IN THE SHADOW OF THE LANDER?

WITH SUCH DEEP SHADOWS, THIS GUY SHOULD BE IN SILHOUETTE.

THERE MUST BE ANOTHER LIGHT SOURCE.

LOOK AT THIS BRIGHT SURFACE.

WE KNOW THIS SURFACE REFLECTS LIGHT, BECAUSE IF IT DIDN'T, WE WOULDN'T BE ABLE TO SEE IT.

THE MOON IS LUMINOUS.

OBJECTS, EVEN IN SHADOW, ARE BATHED IN THIS LIGHT.

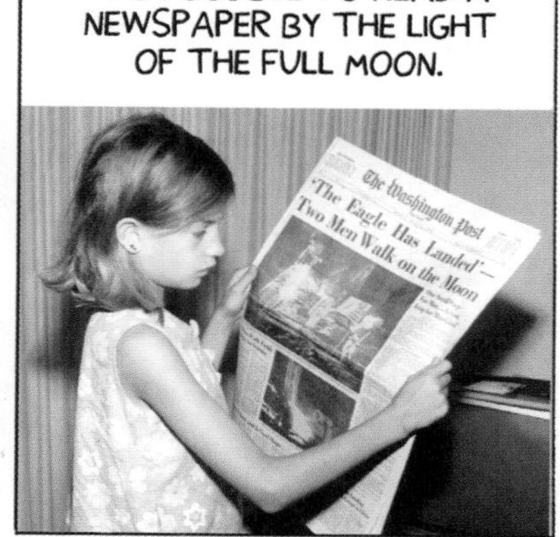
IT'S POSSIBLE TO READ A NEWSPAPER BY THE LIGHT OF THE FULL MOON.
The Washington Post
'The Eagle Has Landed'—
Two Men Walk on the Moon

NO OTHER LIGHT SOURCE IS NEEDED.
The Washington Post
'The Eagle Has Landed'—
Two Men Walk on the Moon

THE UNITED STATES FLAG RIPPLES AND BENDS, AS IF IN A BREEZE.

HOW IS THIS POSSIBLE WHEN THERE'S NO ATMOSPHERE ON THE MOON?

THE FLAG MOVED ONLY WHEN IT WAS BEING ERECTED. THE MOMENT THE FLAG WAS IN POSITION, IT STOPPED DEAD.

THE FLAG WAS CREATED WITH A RIGID EXTENDABLE SUPPORT PIECE RUNNING ALONG ITS TOP, SO THAT IT WOULD LOOK TAUT.

AT THEIR TECHNICAL DEBRIEFING, THE ASTRONAUTS REPORTED A FEW PROBLEMS WITH THE FLAG DEPLOYMENT.

THEY HAD TROUBLE EXTENDING THE HORIZONTAL TELESCOPING ROD AND COULD NOT PULL IT ALL THE WAY OUT.

THIS GAVE THE FLAG A RIPPLE EFFECT. LATER CREWS INTENTIONALLY LEFT THE ROD PARTIALLY RETRACTED, AS THEY LIKED THE WAY IT LOOKED.

A ROCKET CAPABLE OF LANDING ON THE MOON SHOULD HAVE BURNED A HUGE CRATER ON THE SURFACE.

YET THERE'S NOTHING THERE.

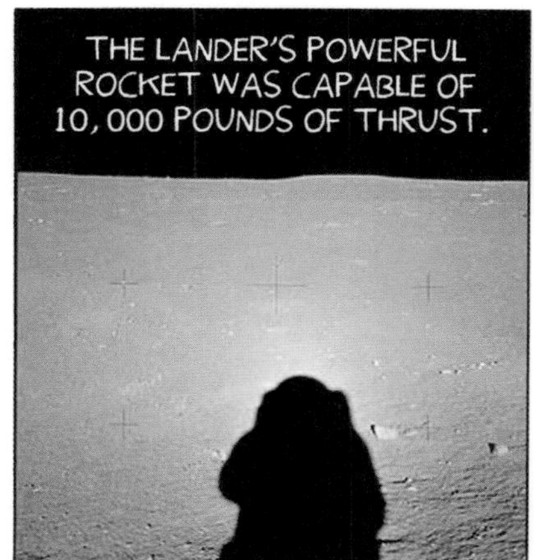
THE LANDER'S POWERFUL ROCKET WAS CAPABLE OF 10,000 POUNDS OF THRUST.

HOWEVER, A REDUCED 3,000 WAS ALL THAT WAS NEEDED TO ENSURE A SAFE LANDING.

FURTHERMORE, THERE'S NO ATMOSPHERE ON THE MOON, AND SO NO AIR PRESSURE.

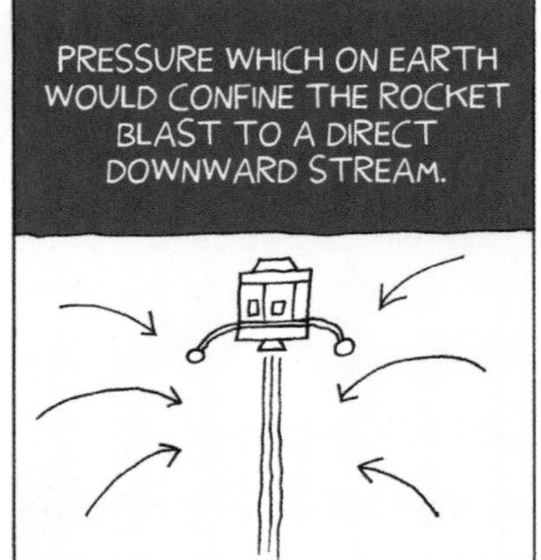
PRESSURE WHICH ON EARTH WOULD CONFINE THE ROCKET BLAST TO A DIRECT DOWNWARD STREAM.

ON THE MOON, THE ROCKET BLAST WAS FREE TO SPREAD OUT OVER A MUCH WIDER AREA.

THIS BLEW A BIT OF DUST AROUND, BUT NOT MUCH ELSE.

WHY ARE THERE NO VISIBLE FLAMES FROM THE ROCKET?

THE FUEL USED (HYDRAZINE AND DINITROGEN TETROXIDE) MIXES TOGETHER AND IGNITES INSTANTLY.

THIS PRODUCES A FLAME THAT IS COMPLETELY TRANSPARENT.

THIS FUEL IS STILL USED IN SOME ORBITING SPACE CRAFT AND IMAGES OF THESE CRAFT SHOW NO FLAMES.

IF YOU RUN THE MOON FOOTAGE AT DOUBLE SPEED, IT LOOKS LIKE IT WAS FILMED ON EARTH.

IN 2005, JAMIE HYNEMAN AND ADAM SAVAGE OF THE POPULAR DISCOVERY CHANNEL SHOW MYTHBUSTERS...

TESTED MANY OF THE MOON HOAX CLAIMS.

THEY ATTEMPTED TO DUPLICATE THE ASTRONAUT'S MOON WALK IN EARTH'S GRAVITY.

A HARNESS WAS USED TO MAKE THE WEARER ONE SIXTH OF HIS NORMAL WEIGHT.
WOO!

THE WEARER'S MOVEMENTS WERE THEN SHOT AT 48 FRAMES A SECOND.

THE TAPE WAS THEN PLAYED BACK AT THE REGULAR 24 PER SECOND.

THE SLOW MOTION RESULTS, ALTHOUGH QUITE CLOSE TO THE NASA FOOTAGE...

STILL LACKED THE SMOOTH LOW-GRAVITY LOOK OF THE ASTRONAUT'S WALK.

THE MYTHBUSTERS THEN WENT ON TO COMPARE THIS RESULT WITH AN ACTUAL DEMONSTRATION IN MICRO-GRAVITY.

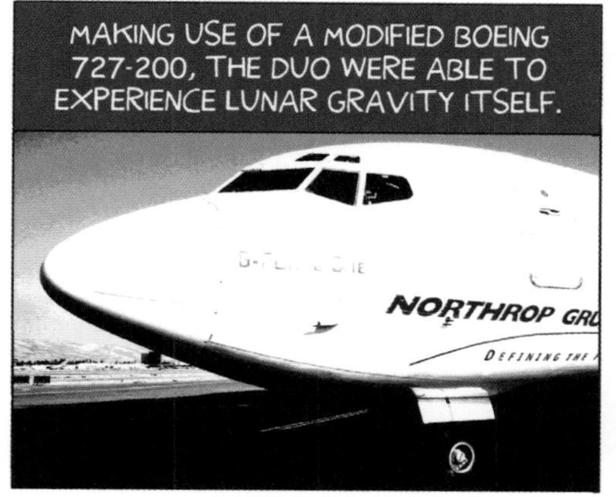
MAKING USE OF A MODIFIED BOEING 727-200, THE DUO WERE ABLE TO EXPERIENCE LUNAR GRAVITY ITSELF.
NORTHROP GRU

THIS WAS ACHIEVED BY HAVING THE AIRCRAFT FLY A SERIES OF PARABOLIC ARCS.

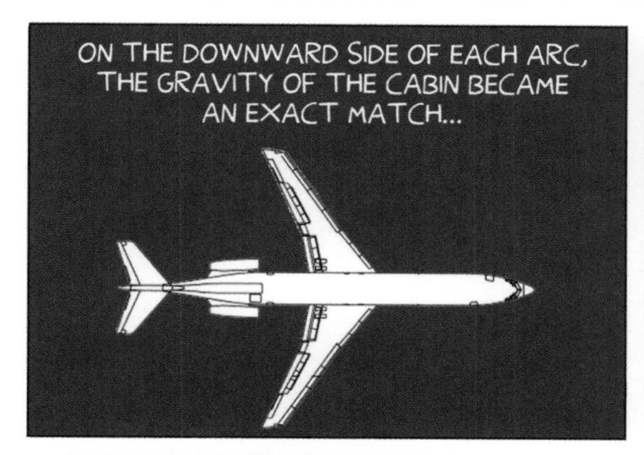
ON THE DOWNWARD SIDE OF EACH ARC, THE GRAVITY OF THE CABIN BECAME AN EXACT MATCH...

WITH THE MOON'S GRAVITATIONAL PULL.

IN THE SHOW ADAM SAVAGE TALKS ABOUT HIS EXPERIENCE OF WALKING IN MICRO-GRAVITY.

ALL THE NASA FOOTAGE MAKES SENSE TO ME NOW.

COUNT DOWN

Wapakoneta Daily News

NEIL STEPS ON THE MOON

"All people one," Nixon tells pair

Armstrong, Aldrin set for hazardous return

Navy beefs up armada in Pacific

Mom, dad worries lighter after walk

THESE ROCKS, CLEARLY FORMED IN THE ABSENCE OF OXYGEN OR WATER, COULD NOT HAVE BEEN COLLECTED OR MANUFACTURED ON EARTH.

THE OLDEST MOON ROCKS ARE UP TO 4.5 BILLION YEARS OLD...

MAKING THEM 200 MILLION YEARS OLDER THAN THE OLDEST EARTH ROCKS.

MOON ROCKS ARE UNIQUE. LUNAR SAMPLES HAVE ALMOST NO WATER TRAPPED IN THEIR CRYSTAL STRUCTURE.

AND COMMON EARTH SUBSTANCES, SUCH AS CLAY MINERALS, ARE ABSENT IN MOON ROCK.

LOOK CLOSE AND YOU'LL SEE THAT APOLLO MOON ROCKS ARE PEPPERED WITH TINY CRATERS FROM METEOROID IMPACTS.

THIS COULD ONLY HAPPEN TO ROCKS FROM A PLANET WITH LITTLE OR OR NO ATMOSPHERE...

LIKE THE MOON.

I COULD GO ON REBUTTING THE MOON HOAX CLAIMS FOR HUNDREDS OF PAGES.

BUT I KNOW IT WON'T MAKE ANY DIFFERENCE TO THE HARD-CORE BELIEVERS.

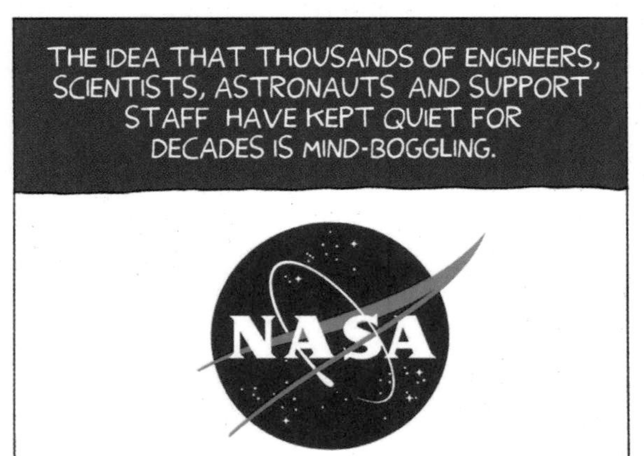
THE IDEA THAT THOUSANDS OF ENGINEERS, SCIENTISTS, ASTRONAUTS AND SUPPORT STAFF HAVE KEPT QUIET FOR DECADES IS MIND-BOGGLING.
NASA

BUT THERE WILL ALWAYS BE PEOPLE WHO'LL CLAIM A CONSPIRACY FOR JUST ABOUT ANYTHING.
APOLLO 11

REAL CONSPIRACIES ARE MESSY, FLAWED, HUMAN AFFAIRS.

HISTORY HAS SHOWN US REPEATEDLY HOW DIFFICULT IT IS EVEN FOR THE MOST POWERFUL TO KEEP THEIR DUBIOUS ACTIVITIES SECRET.
The New York Times
NIXON RESIGNS
HE URGES A TIME OF 'HEALING'; FORD WILL TAKE OFFICE TODAY
'Sacrifice' Is Praised; Kissinger to Remain
The 37th President Is First to Quit Post

HOW CREDIBLE IS IT, THEN, THAT IN THE WORLD OF THE INTERNET, OF TALK SHOWS, FILM AND BOOK DEALS...

NOT ONE ASTRONAUT OR NASA EMPLOYEE HAS BLOWN THE WHISTLE ON THIS ALLEGED HOAX?

IN ORDER TO FOOL BOTH THE WORLD, AND ITS OWN PEOPLE, THE UNITED STATES GOVERNMENT WOULD HAVE HAD TO BUILD A GLOBAL BILLION DOLLAR INFRASTRUCTURE.

FAR EASIER TO ACTUALLY GO TO THE MOON. WHICH IS, OF COURSE, WHAT THEY DID.
END

FRACKING

FRACKING IS A SLANG TERM
FOR HYDRAULIC FRACTURING.

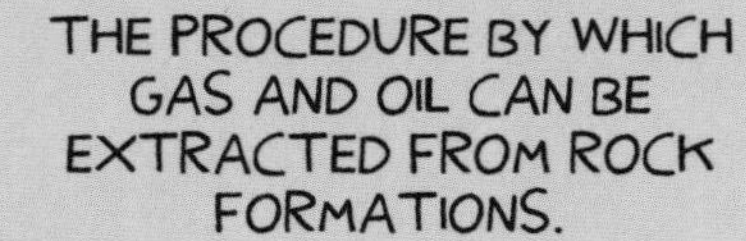
THE PROCEDURE BY WHICH
GAS AND OIL CAN BE
EXTRACTED FROM ROCK
FORMATIONS.

FLUID IS BLASTED AT HIGH
PRESSURE INTO CRACKS IN
THE ROCK, CAUSING
FURTHER FRACTURING.

THESE LARGER FISSURES
ALLOW MORE OIL AND GAS
OUT OF THE WELL BORE...

FROM WHERE IT CAN
BE REMOVED.

SOUNDS CLEAN AND
EFFICIENT, DOESN'T IT?

THE SHALE DEPOSITS THAT CONTAIN THE GAS ARE OFTEN A MILE OR MORE BELOW THE SURFACE.

THIS IS WAY BELOW ANY UNDERGROUND SOURCE OF DRINKING WATER.

SEVERAL LAYERS OF STEEL CASING ENCLOSE THE WELL BORE...

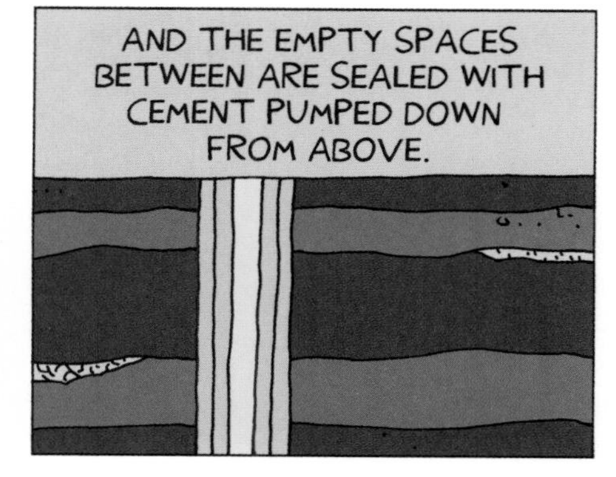
AND THE EMPTY SPACES BETWEEN ARE SEALED WITH CEMENT PUMPED DOWN FROM ABOVE.

HOWEVER, WHAT MAY LOOK STURDY IN ANY DIAGRAM CAN BE FRAGILE IN REALITY. (FOR A START, BECAUSE OF THE LENGTH OF THE WELL.)

IT'S THOUGHT THAT MUCH OF THE CONCRETE CAN BE LOST WHEN THE CASING IS BEING CEMENTED.

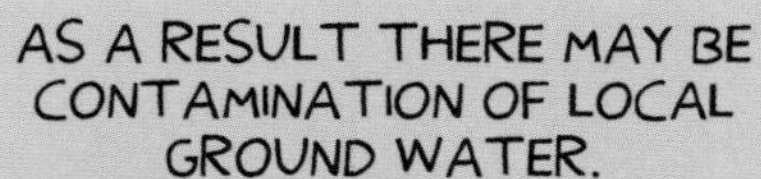

NOT JUST GAS MIGRATION, BUT ALSO TOXIC CHEMICALS.

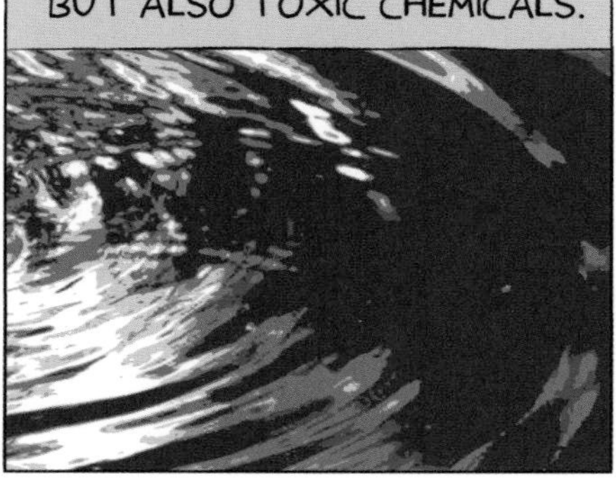

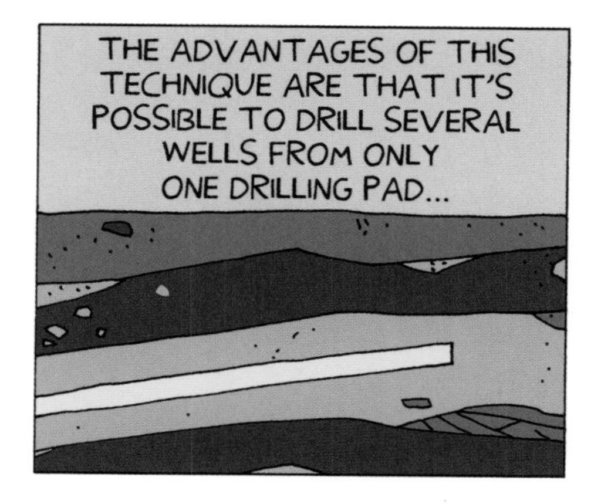
THE ADVANTAGES OF THIS TECHNIQUE ARE THAT IT'S POSSIBLE TO DRILL SEVERAL WELLS FROM ONLY ONE DRILLING PAD...

AND SO MINIMISE THE IMPACT ON THE SURFACE ENVIRONMENT.

ONCE THE REQUIRED LENGTH IS DRILLED OUT, THE DRILL IS REMOVED...

AND A PERFORATING GUN CONTAINING EXPLOSIVES IS LOWERED DOWN INTO THE ROCK LAYER.

WHEN THE GUN IS FIRED IT CREATES HOLES THROUGH THE CASING, CEMENT, AND INTO THE TARGET ROCK.

THESE VERY SMALL EXPLOSIONS CREATE PERFORATIONS ONLY A FEW INCHES LONG.

THE PERFORATION GUN IS THEN REMOVED IN PREPARATION FOR THE NEXT STEP.

THE HYDRAULIC FRACTURING ITSELF.

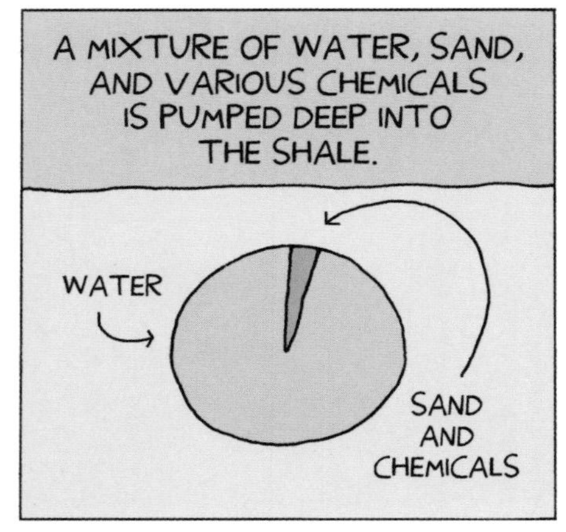
A MIXTURE OF WATER, SAND, AND VARIOUS CHEMICALS IS PUMPED DEEP INTO THE SHALE.
WATER
SAND AND CHEMICALS

THE CHEMICALS BOTH HELP LUBRICATE THE PROCESS AND KEEP BACTERIA FORMING.
WATER
SAND AND CHEMICALS

THE FORCE OF THIS LIQUID THROUGH THE PERFORATIONS FRACTURES THE SHALE.
CRACK
CRACK

PARTICLES OF SAND, CARRIED ALONG WITH THE FLOW, WEDGE IN THE CRACKS, KEEPING THE FISSURES OPEN.

THE FRACTURING IS REPEATED ALONG THE ENTIRE HORIZONTAL LENGTH OF THE WELL, WHICH CAN EXTEND FOR SEVERAL MILES.

AT THIS POINT, AS MUCH OF THE FLUID AS POSSIBLE IS PUMPED OUT AND THE EXTRACTION OF THE GAS BEGINS.

IT'S AN INSPIRED TECHNIQUE. IN RESEARCHING THIS SUBJECT, I COULDN'T HELP BUT ADMIRE THE INGENUITY INVOLVED.

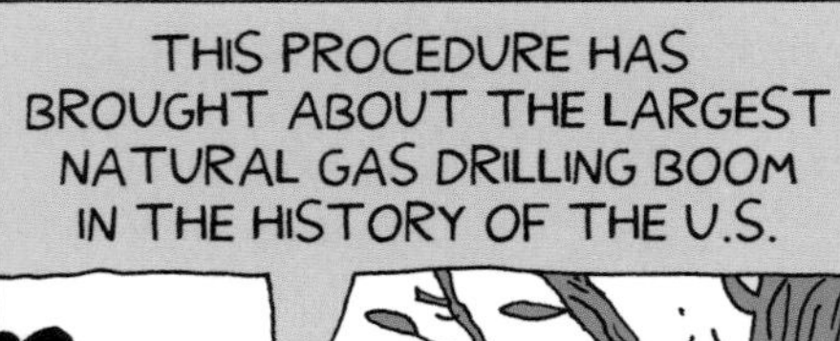
THIS PROCEDURE HAS BROUGHT ABOUT THE LARGEST NATURAL GAS DRILLING BOOM IN THE HISTORY OF THE U.S.

IT HAS ALLOWED ACCESS TO VAST DEPOSITS OF NATURAL GAS, ENOUGH, IT IS CLAIMED, TO SUPPLY THE COUNTRY FOR DECADES.

THE DRILLING BOOM STRETCHES ACROSS 31 STATES AND UP INTO CANADA.

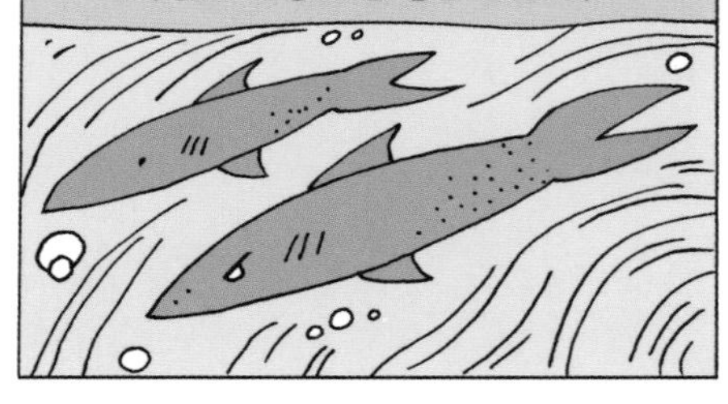
IF IT COULD BE CERTAIN THAT ALL THIS INDUSTRY WOULDN'T HARM THE LANDSCAPE, THEN ALL WOULD BE WELL.

HOWEVER, THERE ARE SERIOUS ENVIRONMENTAL CONCERNS TO CONSIDER.

THESE CONCERNS INCLUDE GROUNDWATER CONTAMINATION...

THROUGH THE MIGRATION OF GASES AND FRACTURING CHEMICALS TO THE SURFACE...

THE MISHANDLING OF WASTE, AND THE RISKS TO AIR QUALITY.

THE CHEMICALS USED IN THE FRACKING PROCESS ARE THEMSELVES A SERIOUS DANGER.

THIS IS TRUE EVEN THOUGH THE CHEMICALS MAKE UP ONLY A SMALL AMOUNT OF THE LIQUID USED.

TYPICALLY ONLY 0.5 PERCENT OF THE TOTAL VOLUME OF THE FLUID.

THIS DOESN'T SOUND LIKE MUCH, I KNOW, BUT MILLIONS OF GALLONS OF WATER ARE INJECTED INTO THE GROUND DURING THIS PROCESS.

THIS MEANS THAT THE AMOUNT OF CHEMICAL PER FRACKING OPERATION IS VERY LARGE.

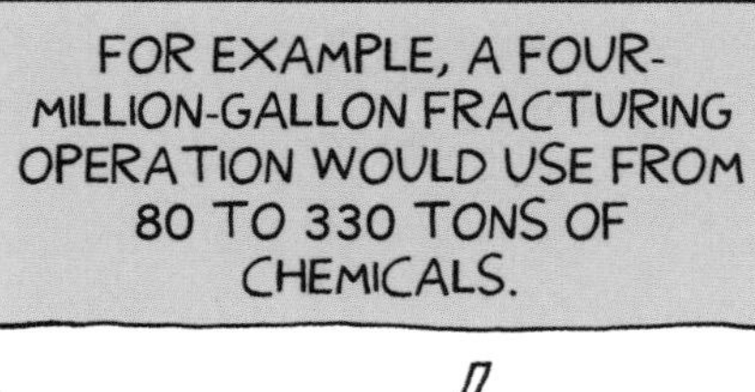
FOR EXAMPLE, A FOUR-MILLION-GALLON FRACTURING OPERATION WOULD USE FROM 80 TO 330 TONS OF CHEMICALS.

HERE ARE SOME OF THE MOST COMMON CHEMICALS USED IN THE FRACKING PROCESS.

METHANOL: FOUND IN ANTI-FREEZE, PAINT SOLVENT, AND VEHICLE FUEL.

BENZENE: COMMONLY FOUND IN GASOLINE. LONG-TIME EXPOSURE CAN CAUSE CANCER, BONE MARROW FAILURE, OR LEUKAEMIA.
YUM!

LEAD: PARTICULARLY HARMFUL TO CHILDREN'S NEUROLOGICAL DEVELOPMENT.
WANT SOME?
YAY!

HYDROGEN FLUORIDE:
FOUND IN RUST REMOVERS,
ALUMINIUM BRIGHTENERS,
AND HEAVY-DUTY CLEANERS.

ITS FUMES ARE HIGHLY
IRRITATING, CORROSIVE,
AND POISONOUS.

NAPHTHALENE: A CARCINOGEN
FOUND IN MOTH BALLS.

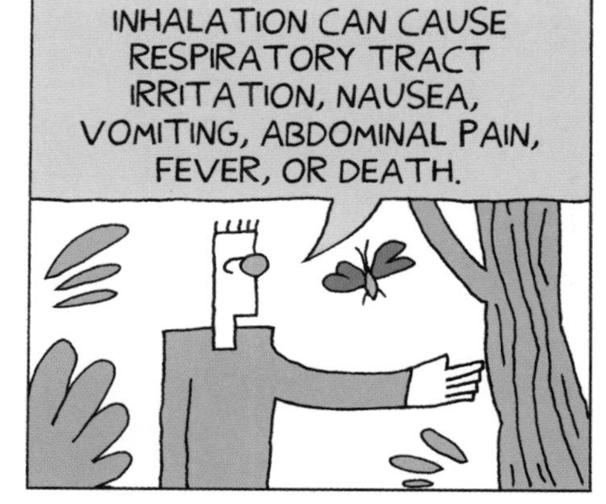
INHALATION CAN CAUSE
RESPIRATORY TRACT
IRRITATION, NAUSEA,
VOMITING, ABDOMINAL PAIN,
FEVER, OR DEATH.

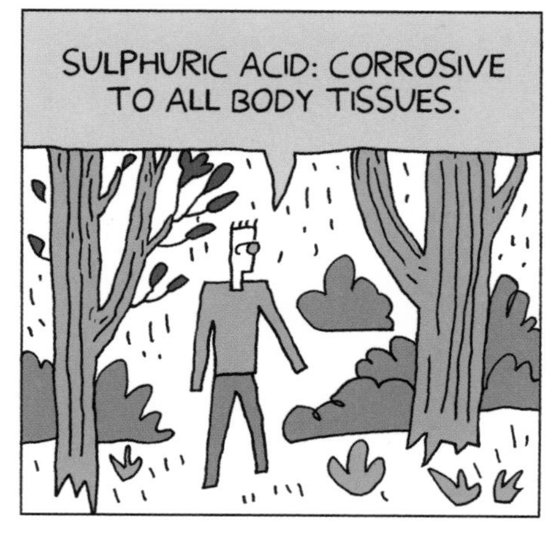
SULPHURIC ACID: CORROSIVE
TO ALL BODY TISSUES.

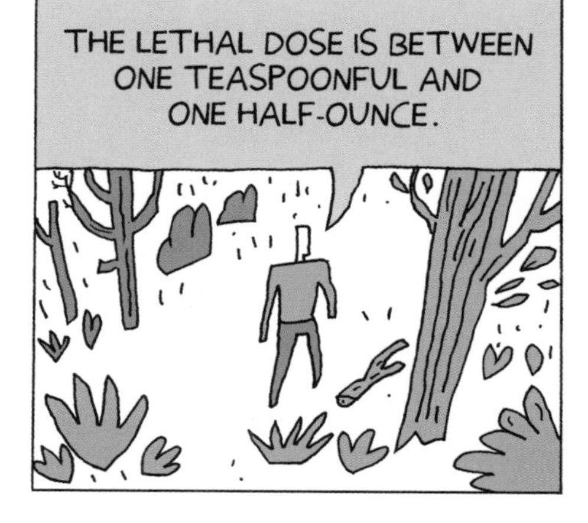
THE LETHAL DOSE IS BETWEEN
ONE TEASPOONFUL AND
ONE HALF-OUNCE.

FORMALDEHYDE: A CARCINOGEN FOUND IN EMBALMING AGENTS.

INGESTION OF EVEN AN OUNCE CAN CAUSE DEATH.

MANY OF THE CHEMICALS USED IN THE FRACKING PROCESS ARE LISTED AS TRADE SECRETS...

WHICH MEANS THAT THEIR EXACT NATURE IS UNKNOWN.

ONLY 25 TO 50 PERCENT OF THIS TOXIC NON-BIODEGRADABLE MATERIAL IS RECOVERED.

THE REST IS SIMPLY LEFT THERE, INFUSED INTO THE LANDSCAPE, FOREVER.

WHAT HAPPENS TO THE LIQUID RECOVERED FROM THE DRILLING OPERATION?
READER'S VOICE

GOOD QUESTION. THE FIRST TYPE OF WASTEWATER IS CALLED 'BRINE'.

THIS BRINE CONTAINS CUTTINGS FROM THE DRILLING PROCESS THAT ARE FULL OF MINERAL SALTS.

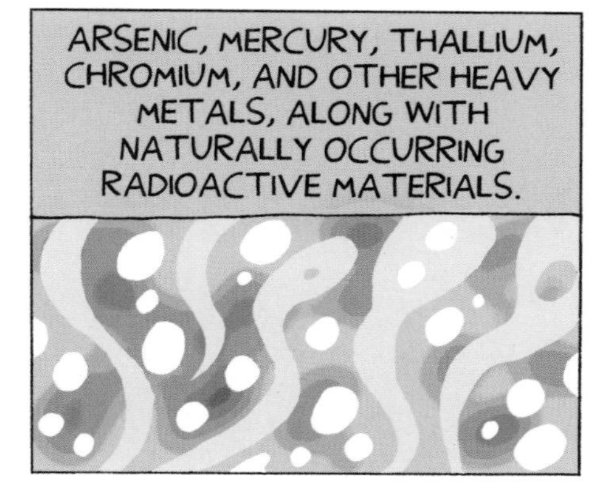
ARSENIC, MERCURY, THALLIUM, CHROMIUM, AND OTHER HEAVY METALS, ALONG WITH NATURALLY OCCURRING RADIOACTIVE MATERIALS.

BRINE IS GENERALLY DUMPED IN HUGE OPEN PITS LINED WITH PLASTIC.

THE FRACKING INDUSTRY SAYS THAT THEY ARE INCREASINGLY USING TANKS TO CONTAIN THIS BRINE.
DANGER
RESTRICTED
AREA

HOWEVER, THERE'S NOT MUCH EVIDENCE OF THIS. PITS ARE COMMON PRACTICE IN MOST STATES.
DANGE
ESTRICT

THE BRINE IS LEFT TO EVAPORATE.
BENZENE FUMES

THE SECOND TYPE OF WASTEWATER IS CALLED 'PRODUCED WATER'.

THIS IS LIQUID THAT CONTINUES TO COME TO THE SURFACE WHEN THE GAS IS BEING EXTRACTED.

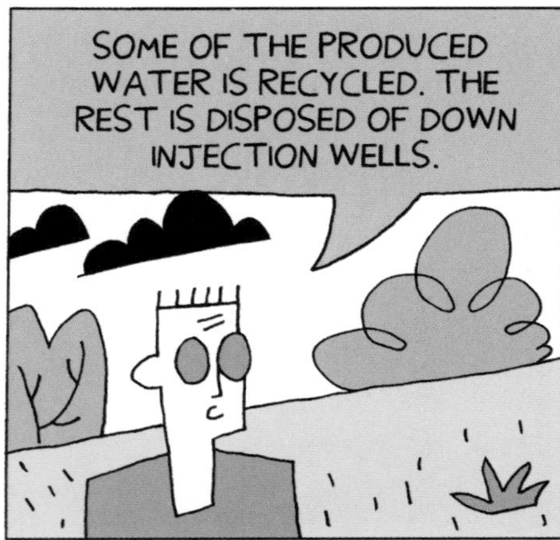
SOME OF THE PRODUCED WATER IS RECYCLED. THE REST IS DISPOSED OF DOWN INJECTION WELLS.

MOST STATES REQUIRE DRILLERS TO DISPOSE OF THIS WATER IN UNDERGROUND STORAGE WELLS BELOW IMPERMEABLE ROCK LAYERS.

THE EXCEPTION BEING PENNSYLVANIA.

IT'S THE ONLY STATE THAT HAS ALLOWED DRILLERS TO DISCHARGE THEIR WASTE THROUGH SEWAGE TREATMENT PLANTS INTO RIVERS.

THE THEORY IS THAT PASSING THIS WASTE THROUGH THE PLANTS IS SAFE BECAUSE MOST TOXIC MATERIAL WILL SETTLE INTO A SLUDGE DURING THE TREATMENT PROCESS.

SLUDGE THAT CAN BE TRUCKED TO A LANDFILL.

AND WHATEVER TOXIC MATERIAL REMAINS WILL BE DILUTED WHEN MIXED INTO RIVERS.

BUT, IN 2008, SOME SEWERAGE PLANTS WERE TAKING SUCH LARGE AMOUNTS OF WASTE WITH HIGH SALT LEVELS THAT...

DOWNSTREAM, UTILITIES BEGAN COMPLAINING THAT THE RIVER WATER WAS EATING AWAY AT THEIR MACHINES.
IT'S BROKEN.

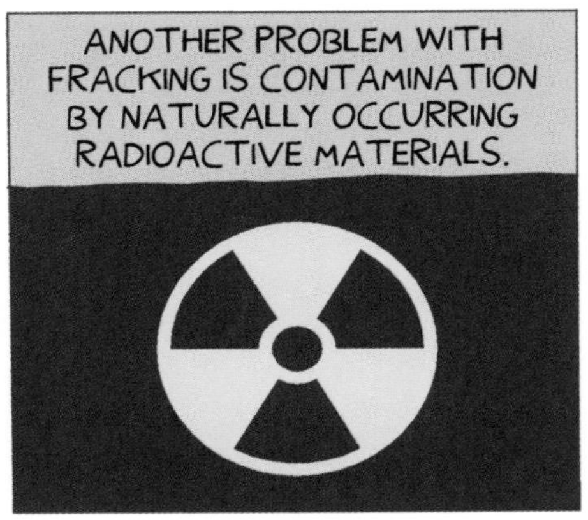
ANOTHER PROBLEM WITH FRACKING IS CONTAMINATION BY NATURALLY OCCURRING RADIOACTIVE MATERIALS.

MOST SEWAGE TREATMENT PLANTS ARE NOT EQUIPPED TO REMOVE THIS MATERIAL...

NOR ARE THEY REQUIRED BY LAW TO TEST FOR IT.

THESE LOW LEVELS OF RADIOACTIVITY POSE NO THREAT TO THE PUBLIC.
INDUSTRY SPOKESMAN

IT'S MORE OF A PERCEPTION ISSUE THAN A REAL HEALTH THREAT.

IN 2011 THE NEW YORK TIMES REPORTED THAT OF MORE THAN 179 WELLS PRODUCING WASTEWATER...

WITH HIGH LEVELS OF RADIATION ...

AT LEAST 116 REPORTED LEVELS OF RADIOACTIVE MATERIAL 100 TIMES GREATER...

THAN THOSE SET BY FEDERAL DRINKING WATER STANDARDS.

AT LEAST 15 OF THESE WELLS PRODUCED WASTEWATER WITH LEVELS OF RADIATION 1000 TIMES HIGHER...

THAN ACCEPTABLE FEDERAL STANDARDS FOR DRINKING WATER.

WE'RE INCREASINGLY RECYCLING WASTEWATER RATHER THAN DISPOSING OF IT.

EVEN WITH RECYCLING, THE AMOUNT OF WASTEWATER PRODUCED IN PENNSYLVANIA IS EXPECTED TO INCREASE.

ACCORDING TO INDUSTRY PROJECTIONS, 50,000 NEW WELLS ARE LIKELY TO BE DRILLED IN THE STATE OVER THE NEXT TWO DECADES.

IF YOU SCALE UP ALL THE GAS DRILLING ACTIVITY ACROSS THE ENTIRE UNITED STATES...

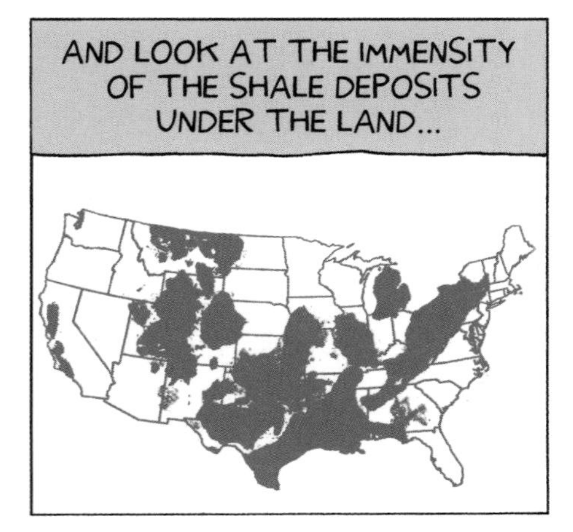
AND LOOK AT THE IMMENSITY OF THE SHALE DEPOSITS UNDER THE LAND...

YOU HAVE TO ASK, HOW CAN THE VAST AMOUNTS OF WASTE BE SAFELY DISPOSED OF?

THE GAS INDUSTRY CLAIMS THERE IS NO POSSIBILITY OF LEAKAGE AND CONTAMINATION...

BECAUSE ANY WASTE MATERIAL IS DEEP BELOW IMPERMEABLE ROCK LAYERS.

THERE IS NO WAY WAY IT CAN RETURN TO THE SURFACE. THERE HAS NEVER BEEN AN EXAMPLE OF THIS HAPPENING.

WELL, I'VE SEEN FOOTAGE OF PEOPLE SETTING FIRE TO THEIR DRINKING WATER...

BECAUSE THE WATER IS SO FULL OF FLAMMABLE METHANE GAS.
CHECK THIS OUT!

WOOAH!

THE TYPE OF METHANE OFTEN REPORTED IN DRINKING WATER IS BIOGENIC IN ORIGIN.

MEANING THAT IT WAS FORMED FROM DECAYING ORGANIC MATTER NEAR THE SURFACE AND HAD NOTHING TO DO WITH GAS DRILLING.

IT'S SIMPLY IMPOSSIBLE FOR THERMOGENIC GAS - THE KIND OF GAS WE DRILL FOR...

TO WORK ITS WAY TO THE SURFACE THROUGH MORE THAN A MILE OF ROCK.

YOU DON'T KNOW THAT BECAUSE NOT NEARLY ENOUGH RESEARCH HAS BEEN DONE TO CONFIRM OR DENY IT.

WE DON'T YET KNOW WHAT THE EFFECT OF SO MANY CLOSELY SPACED, SHALLOW FRACKS COULD BE.

THE PROCESS COULD CREATE ROCK DISTURBANCES

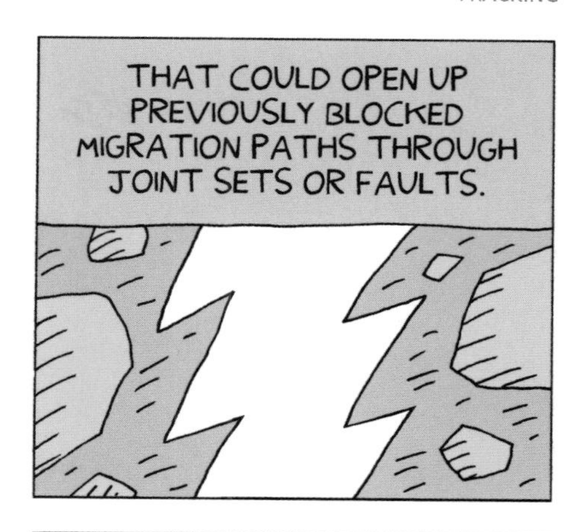
THAT COULD OPEN UP PREVIOUSLY BLOCKED MIGRATION PATHS THROUGH JOINT SETS OR FAULTS.

ALTHOUGH THE ENVIRONMENTAL PROTECTION AGENCY HAS SINCE BACKED AWAY FROM ITS CLAIMS THAT IT HAD FOUND EVIDENCE OF CONTAMINATED GROUNDWATER IN PENNSYLVANIA...

AN INDEPENDENT STUDY BY ENVIRONMENTAL SCIENTISTS AT DUKE UNIVERSITY, DURHAM, NORTH CAROLINA, FOUND CLEAR EVIDENCE OF CONTAMINATION.
GLOOP!
BUBBLE

EVEN IF THE DEEP STORAGE OF THIS MATERIAL DID TURN OUT TO BE SAFE, THERE IS STILL THE ISSUE OF WELL CASING FAILURE TO CONSIDER.

A DRILLING WELL HAS TO BE DEEP, SOMETIMES AS MUCH AS A MILE DEEP AND TWO MILES OUT.

AS WE SAW, THE PIPE IS SHIELDED WITH CEMENT ALONG ITS ENTIRE LENGTH, PUMPED DOWN FROM ABOVE.

THE CEMENT IS SUPPOSED TO SEAL OFF GAS AND CONTAMINANTS FROM LEAKING INTO LOCAL AQUIFERS.

BUT CEMENT FAILURE IN THE OIL AND GAS INDUSTRY DOES HAPPEN.

CEMENT FAILURE CAUSED THE BLOWOUT IN THE GULF OF MEXICO.

THERE DOESN'T NEED TO BE LEAKAGE FROM DEEP UNDERGROUND FOR THERE TO BE CONTAMINATION.

IT CAN LEAK RIGHT OUT OF THE PIPE NEAR THE SURFACE.

ACCIDENTS DO HAPPEN AND FRACKING REGULATION IS LAX.

LOOKING AT PENNSYLVANIA AGAIN. FROM OCTOBER 2008 TO OCTOBER 2010...
MAINE
VERMONT
NEW HAMPSHIRE
MASSACHUSETTS
RHODE ISLAND
CONNECTICUT
NEW YORK
MICHIGAN
PENNSYLVANIA
NEW JERSEY
INDIANA
OHIO
MARYLAND
DELAWARE
WEST VIRGINIA
D.C.

REGULATORS WERE MORE THAN TWICE AS LIKELY TO ISSUE A WRITTEN WARNING THAN TO LEVY A FINE...

FOR ENVIRONMENTAL AND SAFETY VIOLATIONS, ACCORDING TO STATE DATA.

15 COMPANIES WERE FINED FOR DRILLING-RELATED VIOLATIONS IN 2008 AND 2009.

THESE COMPANIES EACH PAID AN AVERAGE OF ABOUT $44,000 IN FINES DURING THIS PERIOD.

THIS WAS LESS THAN HALF OF SOME OF THESE COMPANIES' DAILY PROFITS...

AND A TINY FRACTION OF THE MORE THAN $2 MILLION SOME OF THEM PAID ANNUALLY TO HAUL AND TREAT THE WASTE.

TOM CORBET BECAME GOVERNOR OF PENNSYLVANIA IN 2011.

THIS MAN TOOK MORE CAMPAIGN CONTRIBUTIONS FROM THE GAS INDUSTRY THAN ALL OF HIS COMPETITORS COMBINED.
$

CORBET SAID HE WOULD REOPEN STATE LAND TO NEW DRILLING, REVERSING A DECISION HIS PREDECESSOR HAD MADE.

HE ARGUED AGAINST A PROPOSED GAS EXTRACTION TAX ON THE INDUSTRY.
REGULATION OF THE INDUSTRY HAS BEEN TOO AGGRESSIVE.

I WILL DIRECT THE DEPARTMENT OF ENVIRONMENTAL PROTECTION TO SERVE AS A PARTNER WITH BUSINESS, COMMUNITIES, AND LOCAL GOVERNMENT.

IT SHOULD RETURN TO ITS CORE MISSION OF PROTECTING THE ENVIRONMENT BASED ON SOUND SCIENCE.

WHAT WE SEE IN PENNSYLVANIA IS A MICROCOSM OF WHAT IS HAPPENING ACROSS THE UNITED STATES...

WHERE FOR DECADES ENVIRONMENTAL PROTECTIONS HAVE BEEN ROLLED BACK...

AT THE BEHEST OF THE OIL AND GAS INDUSTRY.

AN INDUSTRY THAT GIVES VAST AMOUNTS OF CAMPAIGN CONTRIBUTIONS TO ITS CHOSEN POLITICIANS...

WHOSE LOYALTY IS TO THEIR WEALTHY DONORS BEFORE THE ELECTORATE.

AMONG THE MANY PROVISIONS OF THE 2005 ENERGY BILL WAS ONE THAT'S BECOME KNOWN AS THE HALLIBURTON LOOPHOLE.

THIS PROVISION WAS INSERTED AT THE BIDDING OF THE THEN VICE PRESIDENT, DICK CHENEY...

WHO WAS, OF COURSE, A FORMER CHIEF EXECUTIVE OF HALLIBURTON - THE WORLD'S LARGEST PROVIDER OF FRACKING SERVICES.

THE HALLIBURTON LOOPHOLE STRIPPED THE ENVIRONMENTAL PROTECTION AGENCY OF ITS AUTHORITY TO REGULATE FRACKING.

IT ALSO CHANGED THE DEFINITION OF THE WORD 'POLLUTANT', SO THAT 'MATERIAL INJECTED INTO A WELL TO FACILITATE THE PRODUCTION OF OIL OR GAS' COULD NO LONGER BE CONSIDERED A POLLUTANT.

THAT THERE SHOULD BE SUCH NAKED INTERFERENCE IN THE POLITICAL PROCESS BY BIG BUSINESS IS CLEARLY WRONG.

IT IS SCIENCE, UNAFFECTED BY POLITICAL BIAS OR EMOTION, THAT SHOULD DECIDE WHETHER FRACKING IS ENVIRONMENTALLY SAFE OR NOT.

THE SITUATION IN THE U.K. IS HARDLY ANY BETTER.

HERE ENERGY COMPANIES ALSO APPEAR TO HAVE UNDUE INFLUENCE ON THE POLITICAL PROCESS.

LORD HOWELL, WHO IS AN ENERGY ADVISER AT THE FOREIGN OFFICE...

IS ALSO PRESIDENT OF THE BRITISH INSTITUTE OF ENERGY ECONOMICS (B.I.E.E.), WHICH IS SPONSORED BY SHELL AND B.P.

THE B.I.E.E. IS BASICALLY A LOBBYING ORGANISATION FOR BIG OIL AND GAS COMPANIES.

LORD HOWELL IS ALSO THE FATHER-IN-LAW OF THE CHANCELLOR OF THE EXCHEQUER, GEORGE OSBORNE.

IN JULY 2012, OSBORNE, AGAINST THE WISHES OF THE ENERGY SECRETARY, ED DAVEY...

CUT WIND ENERGY SUBSIDIES BY TEN PERCENT...

WHILE GIVING A 500-MILLION-POUND TAX BREAK TO OFFSHORE DRILLING.

ANOTHER REPRESENTATIVE OF THE OIL AND GAS INDUSTRY AT THE HEART OF GOVERNMENT IS LORD JOHN BROWNE.

BROWNE IS THE FORMER HEAD OF B.P. AND A 30 PERCENT OWNER OF THE FRACKING COMPANY, CUADRILLA.
GODZILLA?

HE'S NOW AN UNELECTED MEMBER OF THE CABINET OFFICE...
HONI SOIT QUI MAL Y PENSE
DIEU ET MON DROIT

WHERE HE HAS POWER TO APPOINT NON-EXECUTIVE DIRECTORS TO GOVERNMENT DEPARTMENTS.
GO FORTH!
O.K. BOSS

HIS APPOINTMENTS HAVE GONE TO THE TREASURY, THE DEPARTMENT OF ENERGY AND CLIMATE CHANGE...

AND THE DEPARTMENT FOR ENVIRONMENT, FOOD AND RURAL AFFAIRS.

IN THE SPRING OF 2010, A SERIES OF EARTH TREMORS STRUCK THE EAST COAST OF LANCASHIRE AND CUMBRIA.
HECK!
RUMBLE

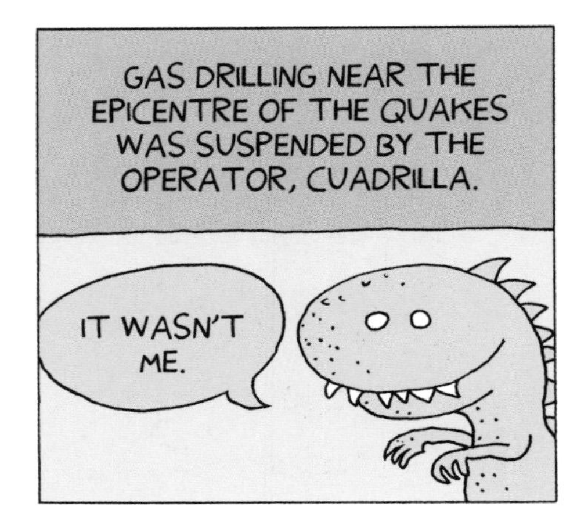
GAS DRILLING NEAR THE EPICENTRE OF THE QUAKES WAS SUSPENDED BY THE OPERATOR, CUADRILLA.
IT WASN'T ME.

ONE OF THE WELLS WAS DEFORMED BY THE SEISMIC ACTIVITY.
IT USUALLY IS YOU.

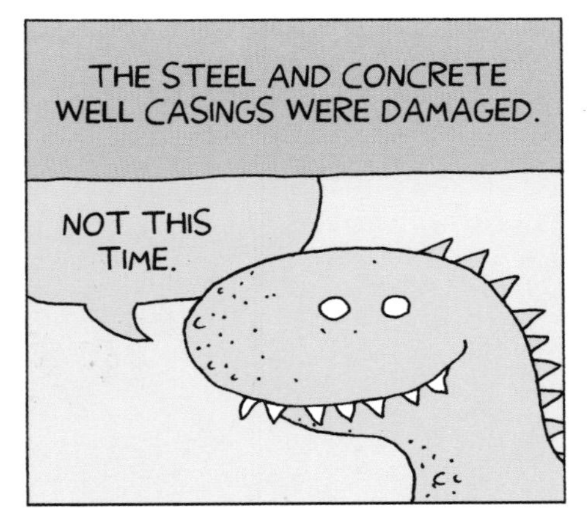
THE STEEL AND CONCRETE WELL CASINGS WERE DAMAGED.
NOT THIS TIME.

AS A RESULT, THERE WAS A DANGER THERE COULD BE A LEAKAGE OF METHANE AND FRACKING FLUIDS.

THE DEPARTMENT OF ENERGY AND CLIMATE CHANGE ASKED A PANEL OF EXPERTS TO REVIEW THE SAFETY OF FRACKING.

THIS PANEL CONSISTED OF PROFESSOR PETER STYLES OF KEELE UNIVERSITY, DR. BRIAN BAPTIE, HEAD OF SEISMOLOGY AT THE BRITISH GEOLOGICAL SURVEY...

AND DR. CHRISTOPHER GREEN, DIRECTOR OF G FRAC TECHNOLOGIES LTD.

CUADRILLA HAD ALREADY ADMITTED THAT ITS PREESE HALL WELL, NEAR BLACKPOOL...

WAS RESPONSIBLE FOR THE MAGNITUDE 2.3 AND 1.5 TREMORS.

THE PANEL RECOMMENDED THAT THE MINING CONTINUED, BUT ONLY WITH TIGHT RESTRICTIONS ON THE PROCESS.

WHICH IS ALL VERY WELL AND GOOD, BUT CAN THE GAS INDUSTRY BE TRUSTED TO ADHERE TO STRICT SAFETY STANDARDS?

THE BEHAVIOUR OF THE INDUSTRY IN THE UNITED STATES SUGGESTS OTHERWISE.

FOR SAFETY POLICIES TO WORK, A STRONG, INDEPENDENT BODY IS NECESSARY TO ENFORCE THE RULES.

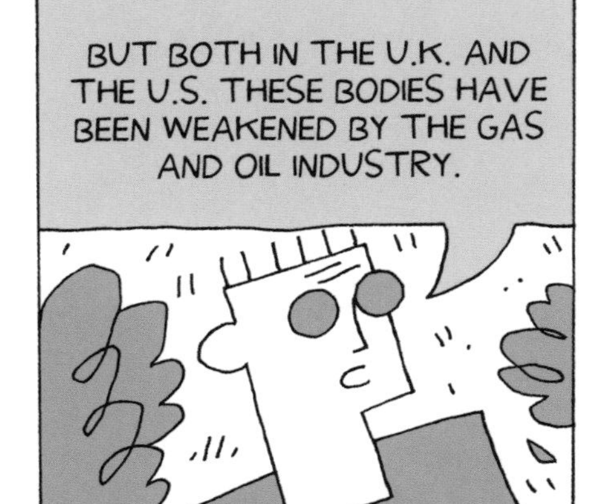
BUT BOTH IN THE U.K. AND THE U.S. THESE BODIES HAVE BEEN WEAKENED BY THE GAS AND OIL INDUSTRY.

IF REGULATORS HAD EASY ACCESS TO INFORMATION AND SUFFICIENT MANPOWER TO ASSESS RISKS...

THEN THE GENERAL PUBLIC WOULD HAVE A CLEARER UNDERSTANDING OF THE BENEFITS OF FRACKING...

COMPARED TO ITS ENVIRONMENTAL COSTS.

AS IT IS, SCIENCE IS AT A CLEAR DISADVANTAGE WHEN FACED WITH THE BEHEMOTH THAT IS THE GAS INDUSTRY.

THIS IS ESPECIALLY TRUE IN THE UNITED STATES, WHERE THE INDUSTRY, WITH ALL ITS POLITICAL AND LEGAL MUSCLE...

AND ITS VAST FINANCIAL RESOURCES, HAS POISONED DEBATE ON THE SUBJECT...

SO THAT ANY CRITICISM OF FRACKING, HOWEVER SLIGHT, LOOKS LIKE LIBERAL BIAS.

HOWEVER, THE INDUSTRY IS NOT ALL-POWERFUL, BECAUSE, AS THE TOBACCO COMPANIES FOUND OUT SOME YEARS AGO...
KAFF! KAFF!

THE TRUTH WILL EVENTUALLY SURFACE.

NO ONE CAN BURY THE TRUTH FOREVER.
END

CLIMATE CHANGE

IN ENGLAND, WHERE I LIVE, THE WINTER OF 2009-2010 WAS UNUSUALLY COLD.

IN FACT, THE NORTHERN HALF OF EUROPE EXPERIENCED ITS COLDEST WINTER SINCE 1981-1982.

THIS, FOR MANY, SHOWED THAT GLOBAL WARMING WAS NONSENSE.

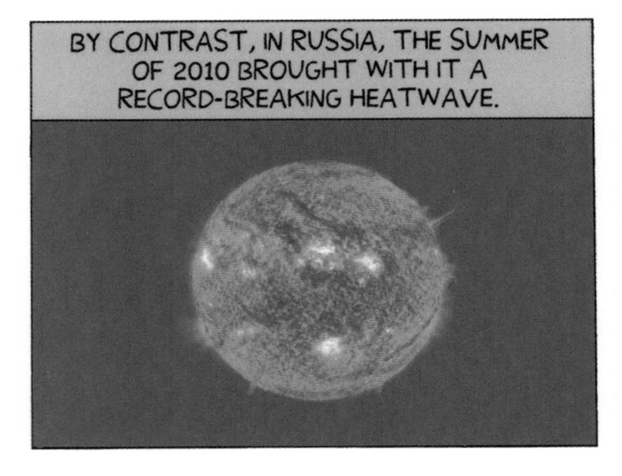
BY CONTRAST, IN RUSSIA, THE SUMMER OF 2010 BROUGHT WITH IT A RECORD-BREAKING HEATWAVE.

FOREST FIRES AND DROUGHT RAVAGED THE COUNTRY.

THE FORMER HAS BEEN CITED AS EVIDENCE THAT CLIMATE CHANGE IS NONSENSE.

WHILE THE LATTER HAS BEEN CITED AS EVIDENCE THAT CLIMATE CHANGE IS REAL.

THE TRUTH IS THAT NEITHER OF THESE EVENTS CAN BE USED TO PROVE THE CASE EITHER WAY.

HUGE AS THESE EXTREME WEATHER EVENTS MAY APPEAR TO US ON OUR TINY HUMAN SCALE...

THEY ARE STILL MERE LOCAL EVENTS. IN ORDER TO GET ANY REAL UNDERSTANDING OF THE PLANET'S CLIMATE...

THEN YOU HAVE TO LOOK AT WEATHER SYSTEMS GLOBALLY OVER A LONG PERIOD OF TIME.

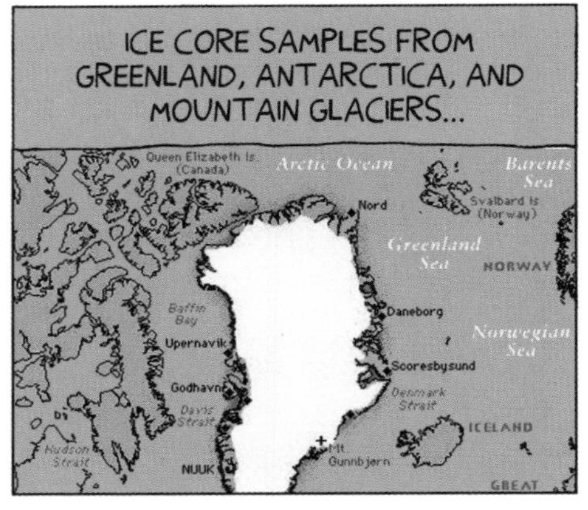
ICE CORE SAMPLES FROM GREENLAND, ANTARCTICA, AND MOUNTAIN GLACIERS...
Queen Elizabeth Is. (Canada)
Arctic Ocean
Barents Sea
Nord
Svalbard Is (Norway)
Greenland Sea
NORWAY
Baffin Bay
Daneborg
Norwegian Sea
Upernavik
Scoresbysund
Godhavn
Denmark Strait
Davis Strait
ICELAND
Hudson Strait
Mt. Gunnbjørn
NUUK
GREAT

HAVE GIVEN SCIENTISTS INFORMATION ON THE EARTH'S CLIMATE GOING BACK THOUSANDS OF YEARS.

IN THE LAST 650, 000 YEARS THERE HAVE BEEN SEVEN CYCLES OF GLACIAL ADVANCE AND RETREAT.

MOST OF THESE CHANGES WERE DUE TO SMALL SHIFTS IN THE EARTH'S ORBIT.

INFORMATION ON THE PRESENT STATE OF THE CLIMATE COMES FROM EARTH-ORBITING SATELLITES AND OTHER TECHNOLOGICAL ADVANCES.

FROM THIS INFORMATION THE EVIDENCE FOR RAPID CLIMATE CHANGE IN MODERN TIMES IS COMPELLING.

GLOBAL SEA LEVELS HAVE RISEN ABOUT 17 CENTIMETRES IN THE PAST CENTURY.

A RATE OF INCREASE THAT HAS DOUBLED IN THE PAST DECADE.

THERE HAS BEEN A CONSISTENT GLOBAL SURFACE TEMPERATURE RISE SINCE THE 1880s...

AND MOST OF THIS WARMING HAS OCCURRED SINCE THE 1970s...

WITH TWO OF THE WARMEST YEARS HAPPENING IN THE PAST 12 YEARS.

ALL THIS HAS TAKEN PLACE EVEN THOUGH THE 2000s HAVE EXPERIENCED A SOLAR OUTPUT DECLINE.

THE OCEANS HAVE ABSORBED MUCH OF THE INCREASED HEAT, WITH THE TOP 700 METRES (ABOUT 2,300 FEET)...

SHOWING WARMING OF 0.302 DEGREES FAHRENHEIT SINCE 1969.

BOTH THE GREENLAND AND ANTARCTIC ICE SHEETS HAVE DECREASED IN MASS.

YES, THAT'S RIGHT! DATA FROM NASA'S GRAVITY RECOVERY AND CLIMATE EXPERIMENT SHOWS THAT...

GREENLAND LOST 150 TO 250 CUBIC KILOMETRES (36 TO 60 CUBIC MILES) OF ICE PER YEAR BETWEEN 2002 AND 2006.

WHILE ANTARCTICA LOST ABOUT 152 CUBIC KILOMETRES (36 CUBIC MILES) OF ICE BETWEEN 2002 AND 2005.

IN JULY 2012 AN ICEBERG TWICE THE SIZE OF MANHATTAN BROKE AWAY FROM THE PETERMAN GLACIER IN GREENLAND...

WHILE SEA ICE IN THE ARCTIC SHRANK TO THE SMALLEST EXTENT EVER RECORDED. SATELLITE IMAGES SHOW THAT THE SUMMER MELT HAD REDUCED THE AREA OF FROZEN SEA TO LESS THAN HALF THE AREA TYPICALLY OCCUPIED FOUR DECADES AGO.

THAT THERE'S BEEN SUCH AN OBVIOUS JUMP IN GLOBAL TEMPERATURE DURING INDUSTRIAL TIMES...

DOES SUGGEST THAT THESE CHANGES ARE HUMAN-MADE.

BUT WHAT IS THE SCIENCE BEHIND THIS THEORY?

THE ARGUMENT FOR HUMAN-DRIVEN CLIMATE CHANGE IS AS FOLLOWS.

THE EARTH'S ATMOSPHERE KEEPS THE PLANET WARMER THAN IT WOULD BE IF IT DIDN'T HAVE AN ATMOSPHERE.
IONOSPHERE
MESOSPHERE
OZONE
STRATOSPHERE
TROPOSPHERE

THE MAIN GASES THAT CONTRIBUTE TO THIS EFFECT ARE CARBON DIOXIDE, METHANE, AND WATER VAPOUR.

THEIR ABILITY TO ACT AS GREENHOUSE GASES CAN BE DEMONSTRATED IN A LABORATORY.

SINCE THE INDUSTRIAL REVOLUTION, THE QUANTITY OF THESE GREENHOUSE GASES IN THE ATMOSPHERE HAS INCREASED SHARPLY.

HUMAN ACTIVITY HAS POURED THESE GASES INTO THE ATMOSPHERE.

THE TEMPERATURE OF THE EARTH'S ATMOSPHERE HAS BEEN RISING, AND IT CONTINUES TO RISE.

AND LASTLY, THE INCREASE IN GLOBAL TEMPERATURE CORRELATES WITH THE INCREASE OF GREENHOUSE GASES.

SO YOU SAY, BUT MANY ARE SCEPTICAL OF THIS ARGUMENT.

THAT'S TRUE. THERE ARE TWO KINDS OF PEOPLE WHO DISAGREE WITH THE THEORY OF HUMAN-MADE CLIMATE CHANGE.

THERE ARE THOSE WHO DON'T KNOW ALL THE INFORMATION AND ARE THEREFORE DOUBTFUL OF THE THEORY.

BUT THEY TEND TO BE AWARE OF THE LIMITS OF THEIR KNOWLEDGE...

AND SO REMAIN OPEN TO THE IDEA THAT CLIMATE CHANGE MIGHT BE REAL.

THEN THERE ARE THOSE WHO SIMPLY REJECT CLIMATE CHANGE, NOT ON SCIENTIFIC GROUNDS, BUT ON THE GROUNDS OF IDEOLOGY AND DOGMA.

"THE GREATEST HOAX EVER PERPETRATED ON THE AMERICAN PEOPLE".
JAMES INHOFE
SENATOR FOR OKLAHOMA

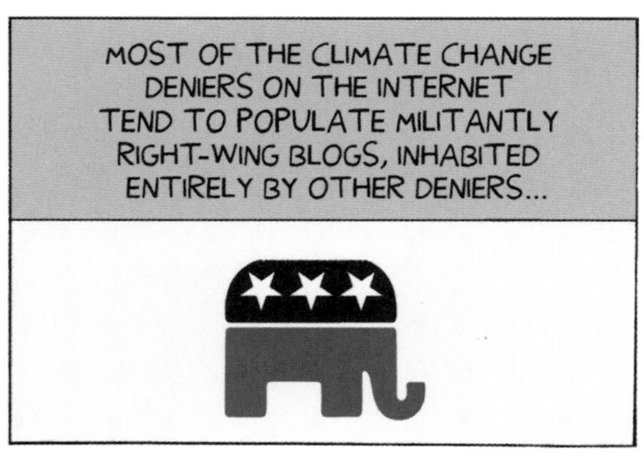
MOST OF THE CLIMATE CHANGE DENIERS ON THE INTERNET TEND TO POPULATE MILITANTLY RIGHT-WING BLOGS, INHABITED ENTIRELY BY OTHER DENIERS...

WHERE A SMATTERING OF CITATIONS FROM LEGITIMATE SCIENTIFIC AUTHORITIES...

ARE USED TO BOLSTER UP THIN ARGUMENTS AND OUTRIGHT DISTORTIONS.
CLIMATE CHANGE HOAX

BUT ISN'T IT TRUE THAT A GROWING NUMBER OF EMINENT SCIENTISTS NOW BELIEVE CLIMATE CHANGE TO BE WRONG?

IN AN ARTICLE PUBLISHED IN THE PROCEEDINGS OF THE NATIONAL ACADEMY OF SCIENCES...

A GROUP OF SCHOLARS DID A STATISTICAL BREAKDOWN OF THE OPINIONS OF THE WORLD'S MOST PROMINENT CLIMATE EXPERTS.
PNAS
Rice pollen and genetic incompatibility

THEIR CONCLUSIONS WERE THAT THOSE SCEPTICAL OF THE EVIDENCE OF HUMAN-MADE CLIMATE CHANGE...

COMPRISED ONLY 2.5 PERCENT OF THE TOP 200 CLIMATE SCIENTISTS.

A TINY SLIVER OF FRINGE OPINION.
30p GET YOUR DAILY EXPRESS FOR JUST 30p
DAILY EXPRESS
Clooney's amazing mother
FREE £5 SPEND AT WHSmith
Iran threatens serious action against sailors
THE BIG CLIMATE CHANGE 'FRAUD'
We are not to blame says top scientist... It's a con to raise tax

IT'S ONE THING TO BE SCEPTICAL, BUT IT'S ANOTHER THING ENTIRELY TO BELIEVE IN A CONSPIRACY.
IPCC
Intergovernmental Panel On Climate Change

"EUROPEANS NOW SEE GLOBAL WARMING AS A MEANS OF HAMPERING U.S. ECONOMIC COMPETITIVENESS"
STEVE MILLOY
FOX NEWS COLUMNIST

A CONSPIRACY THEORY CAN BE DEFINED AS ANY WORLD VIEW THAT TRACES IMPORTANT EVENTS...

TO A SECRETIVE, NEFARIOUS CABAL, AND WHOSE PROPONENTS RESPOND TO CONTRARY FACTS, NOT BY MODIFYING THEIR HYPOTHESES...

BUT INSTEAD BY INSISTING ON THE EXISTENCE OF EVER-WIDENING CIRCLES...

OF HIGH-LEVEL CONSPIRATORS, WHO CONTROL MOST OR ALL PARTS OF SOCIETY.

RADICALISED CLIMATE CHANGE DENIERS BELIEVE THAT GLOBAL WARMING IS A BACK-DOOR METHOD OF CREATING A SOCIALIST STATE.
GREENPEACE

BUT AREN'T YOU ALSO SUGGESTING A CONSPIRACY?

THERE IS REAL OPPOSITION TO THE SCIENTIFIC CONSENSUS, AND THIS OPPOSITION IS CONCENTRATED AROUND POLITICAL MOVEMENTS WITH STRONG TIES TO THE COAL AND OIL INDUSTRY.

EVERY YEAR EXXONMOBIL POURS MILLIONS OF DOLLARS INTO CLIMATE CHANGE DENIAL GROUPS...

THE HEARTLAND INSTITUTE, THE HERITAGE FOUNDATION, THE GEORGE C. MARSHALL INSTITUTE...

THE AMERICAN ENTERPRISE INSTITUTE. ALL GROUPS THAT ARE AT THE HEART OF CLIMATE CHANGE DENIAL.

KOCH INDUSTRIES IS OWNED BY THE BILLIONAIRE KOCH BROTHERS. CHARLES AND DAVID KOCH.

THE COMPANY TURNS OVER ROUGHLY 100 BILLION DOLLARS A YEAR. THE BROTHERS ARE WORTH $21 BILLION EACH.

THE AFP ALSO PROVIDED THE KEY ORGANISING TOOLS THAT SET THE TEA PARTY RUNNING.
1 FOR YOU, 19 FOR ME
TAXED ENOUGH ALREADY
YES WE CAN'T

SOUNDS LIKE ANOTHER CRACK-POT CONSPIRACY THEORY TO ME.

IT'S SIMPLY THE REACTION YOU'D EXPECT FROM BIG BUSINESS WHEN ITS INTERESTS ARE THREATENED.

THE OIL AND COAL INDUSTRIES HAVE A LOT AT STAKE.

IF THE ANSWER TO THE PROBLEM OF CLIMATE CHANGE INVOLVES PHASING OUT FOSSIL FUELS, THEN A DIFFERENT SET OF PEOPLE WILL BE MAKING MONEY.

IT'S NO WONDER THAT THEY'RE FIGHTING TOOTH AND NAIL AGAINST IT.

BUT WHAT ABOUT THIS CLIMATEGATE BUSINESS? WASN'T CLIMATEGATE PROOF THAT ENVIRONMENTALISTS ALSO TWIST THE FACTS TO SUIT THEIR OWN AGENDA?

THE CLIMATEGATE EMAIL CONTROVERSY BEGAN IN NOVEMBER 2009...

WHEN STOLEN EMAILS FROM THE UNIVERSITY OF EAST ANGLIA'S CLIMATE RESEARCH UNIT WERE MADE AVAILABLE TO READ ON THE INTERNET.

OUT OF THESE THOUSANDS OF EMAILS...

A HANDFUL OF STATEMENTS WERE SELECTED, THAT OUT OF CONTEXT COULD BE SEEN AS CONTROVERSIAL...

AS IF SCIENTISTS HAD MANIPULATED RESEARCH IN ORDER TO SUPPORT PRECONCEIVED IDEAS ABOUT CLIMATE CHANGE.

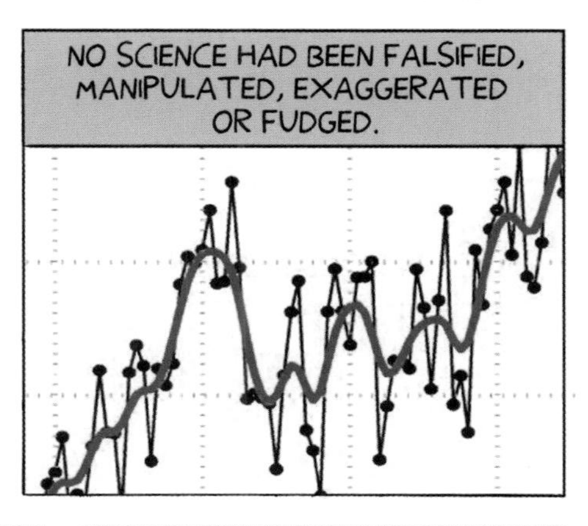

HOWEVER, THE DAMAGE HAD ALREADY BEEN DONE. IN A STUDY, SCOTT MANDIA, A METEOROLOGY PROFESSOR, FOUND THAT MEDIA OUTLETS...

HAD DEVOTED FIVE TO ELEVEN TIMES MORE STORIES TO THE ACCUSATIONS AGAINST THE SCIENTISTS THAN THEY HAD TO THE RESULTING EXONERATIONS.

CLIMATE 'FRAUD'

A CLEAR CASE OF UNFAIR AND UNBALANCED REPORTING.

COMPARE THIS TO THE TINY AMOUNT OF MEDIA COVERAGE THAT KOCH INDUSTRIES RECEIVED...

AFTER THEY WERE CAUGHT RED-HANDED GIVING MONEY TO GROUPS PROMOTING CLIMATE CHANGE AS A HOAX.

SO WHERE DOES THAT LEAVE US?

IT LEAVES US IN A PLACE WHERE BIG OIL CONTINUES TO DISTORT THE DEMOCRATIC PROCESS...

WHILE THE MEDIA LIES ASLEEP AT THE WHEEL.

MEANWHILE THE FUTURE LOOKS BLEAK.

DROUGHT, HUNGER, DISEASE, THE EXTINCTION OF AT LEAST A FOURTH OF THE WORLD'S SPECIES.

THE LOSS OF VITAL GLOBAL NATURAL AREAS LIKE THE AMAZON AND THE GREAT BARRIER REEF.

THE FLOODING OF COASTS AND ISLANDS OCCUPIED BY HUNDREDS OF MILLIONS OF PEOPLE.

THE SCIENCE PREDICTS THAT THESE EVENTS ARE GOING TO HAPPEN. LET'S NOT LEAVE IT TO THE SUPER-RICH TO DECIDE...

WHO LIVES AND WHO DIES.
END

EVOLUTION

HEY, EXPLAIN THIS! WHY ARE THERE STILL APES IF THOSE WERE ALLEGEDLY OUR ANCESTORS?

THAT'S A VERY GOOD QUESTION AND ONE THAT DESERVES AN ANSWER.

THE EXPLANATION IS THAT APES AND HUMANS ARE ON A BRANCH OF THE EVOLUTIONARY TREE...

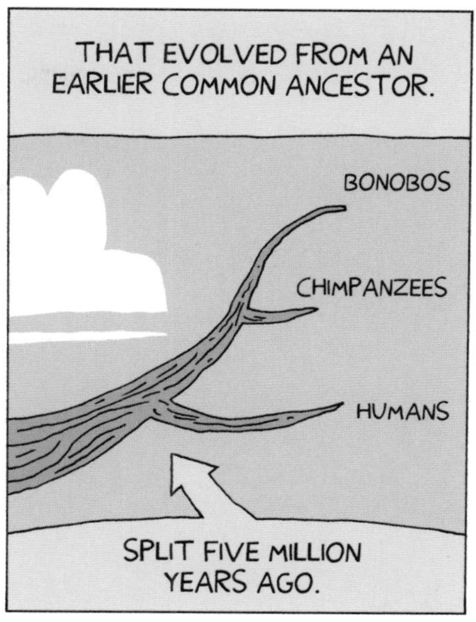
THAT EVOLVED FROM AN EARLIER COMMON ANCESTOR.
BONOBOS
CHIMPANZEES
HUMANS
SPLIT FIVE MILLION YEARS AGO.

THESE SPECIES ARE CONNECTED, NOT FROM ONE TO ANOTHER IN THE PRESENT TIME, BUT IN THE DEEP PAST. IT IS THIS EARLIER COMMON ANCESTOR WHICH IS LONG EXTINCT.
GIBBONS
ORANGUTANS
GORILLAS
BONOBOS
CHIMPANZEES
HUMANS

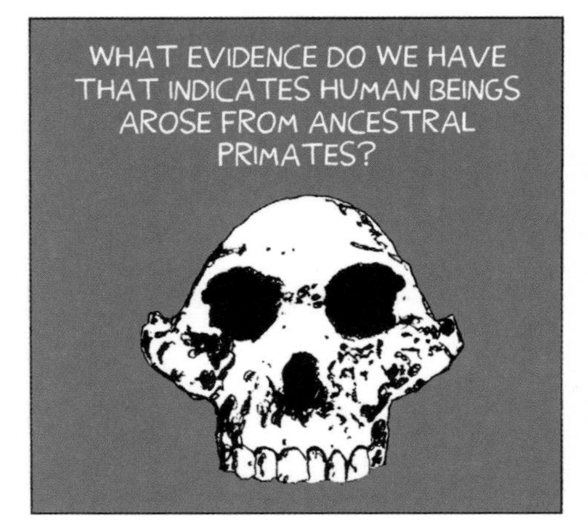
WHAT EVIDENCE DO WE HAVE THAT INDICATES HUMAN BEINGS AROSE FROM ANCESTRAL PRIMATES?

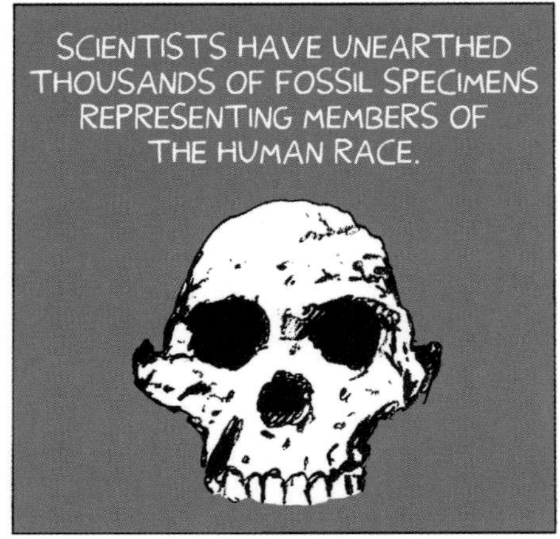
SCIENTISTS HAVE UNEARTHED THOUSANDS OF FOSSIL SPECIMENS REPRESENTING MEMBERS OF THE HUMAN RACE.

THESE FINDS REVEAL A WELL BRANCHED TREE...

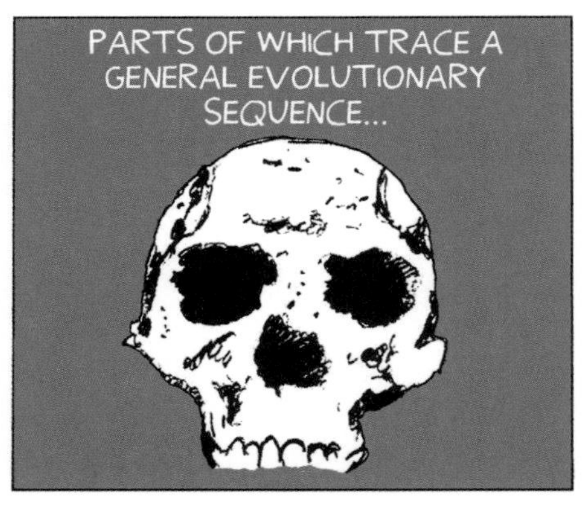
PARTS OF WHICH TRACE A GENERAL EVOLUTIONARY SEQUENCE...

LEADING FROM APE-LIKE FORMS TO MODERN HUMANS.

RECENT SPECIES WITH LARGER BRAINS USED MORE SOPHISTICATED TOOLS THAN THE MORE ANCIENT SPECIES.

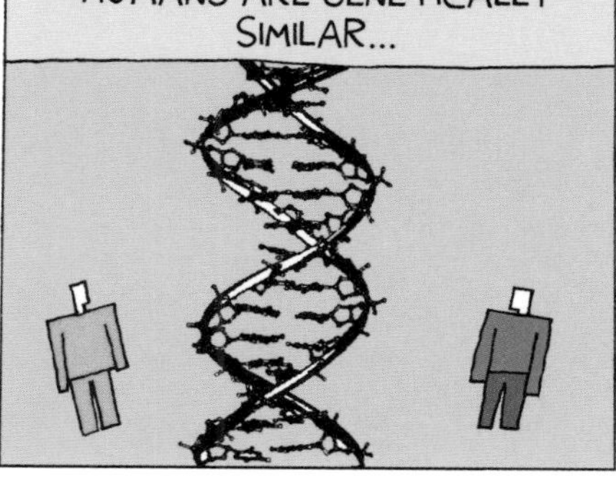

BUT ALL THIS IS JUST A THEORY, ISN'T IT?

HERE ARE SOME EVOLUTIONARY PREDICTIONS.

WE SHOULD BE ABLE TO FIND EVIDENCE FOR EVOLUTIONARY CHANGE IN THE FOSSIL RECORD.

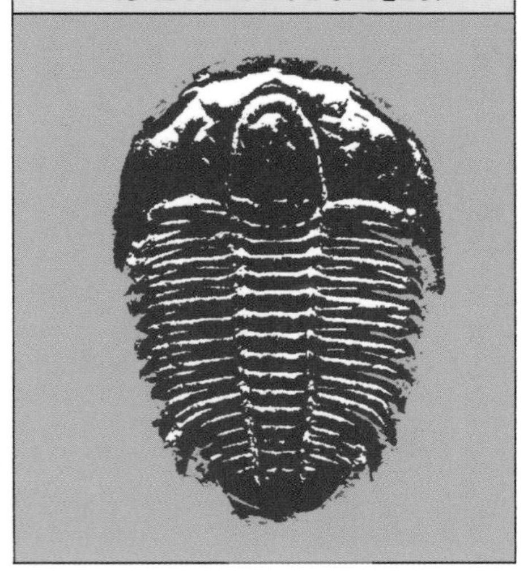
THE DEEPEST AND OLDEST LAYERS OF ROCK WOULD CONTAIN THE FOSSILS OF MORE PRIMITIVE SPECIES.

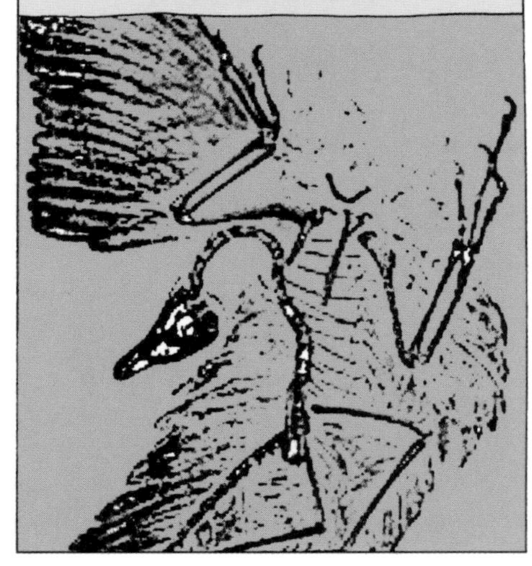
ORGANISMS RESEMBLING PRESENT-DAY SPECIES SHOULD BE FOUND IN THE MOST RECENT LAYERS OF ROCK.

WE SHOULD BE ABLE TO SEE SOME SPECIES
CHANGING OVER TIME, FORMING LINEAGES.

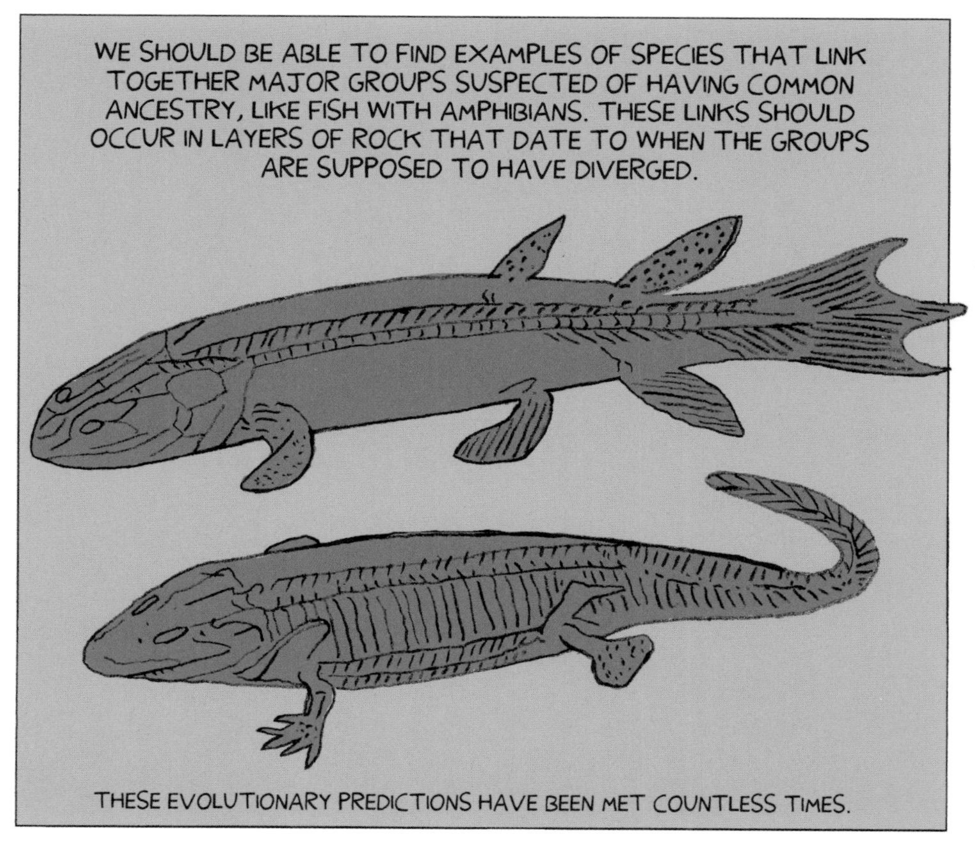
WE SHOULD BE ABLE TO FIND EXAMPLES OF SPECIES THAT LINK TOGETHER MAJOR GROUPS SUSPECTED OF HAVING COMMON ANCESTRY, LIKE FISH WITH AMPHIBIANS. THESE LINKS SHOULD OCCUR IN LAYERS OF ROCK THAT DATE TO WHEN THE GROUPS ARE SUPPOSED TO HAVE DIVERGED.
THESE EVOLUTIONARY PREDICTIONS HAVE BEEN MET COUNTLESS TIMES.

FURTHERMORE, WE SHOULD BE ABLE TO FIND CASES OF IMPERFECT ADAPTATION IN WHICH EVOLUTION...

HAS NOT BEEN ABLE TO ACHIEVE THE SAME DEGREE OF EFFICIENCY AS A CREATOR WOULD.

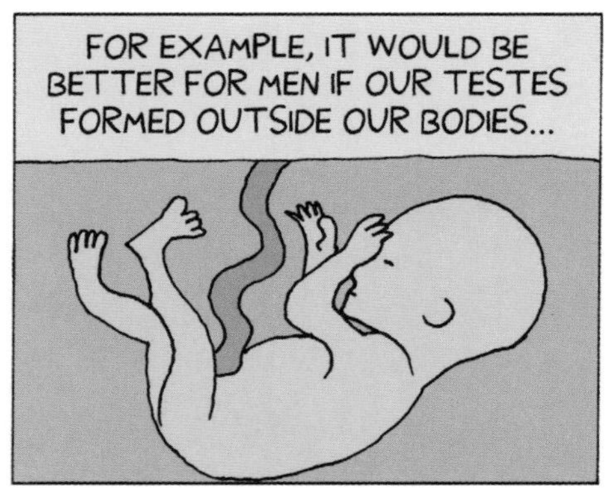
FOR EXAMPLE, IT WOULD BE BETTER FOR MEN IF OUR TESTES FORMED OUTSIDE OUR BODIES...

WHERE THE COOLER TEMPERATURE IS BETTER FOR SPERM.

HOWEVER, THE TESTES BEGIN DEVELOPMENT IN THE ABDOMEN. WHEN THE FOETUS IS SIX OR SEVEN MONTHS OLD...

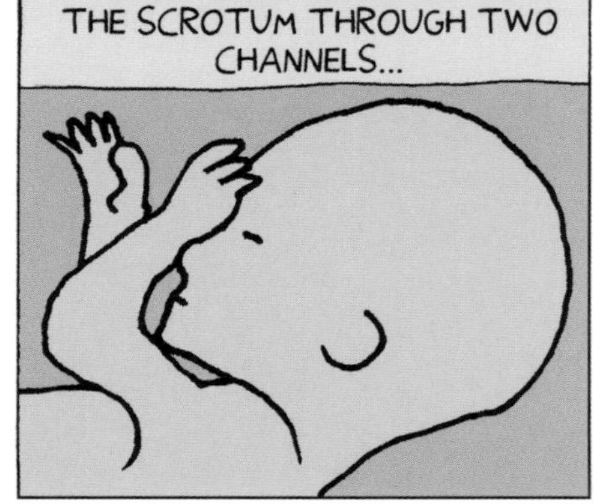
THEY MIGRATE DOWN INTO THE SCROTUM THROUGH TWO CHANNELS...

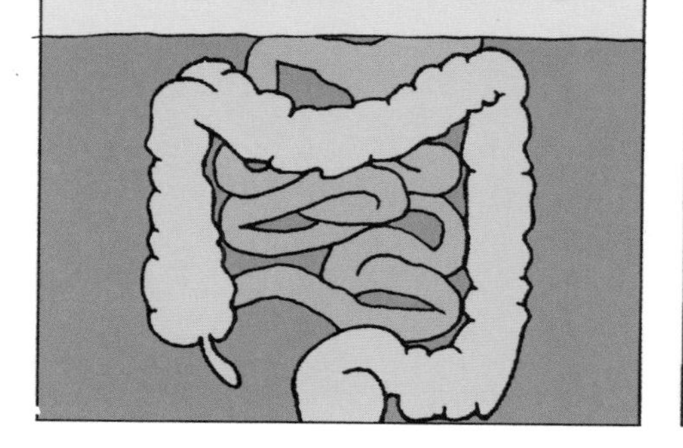
CALLED THE INGUINAL CANALS, REMOVING THEM FROM THE DAMAGING HEAT OF THE BODY.

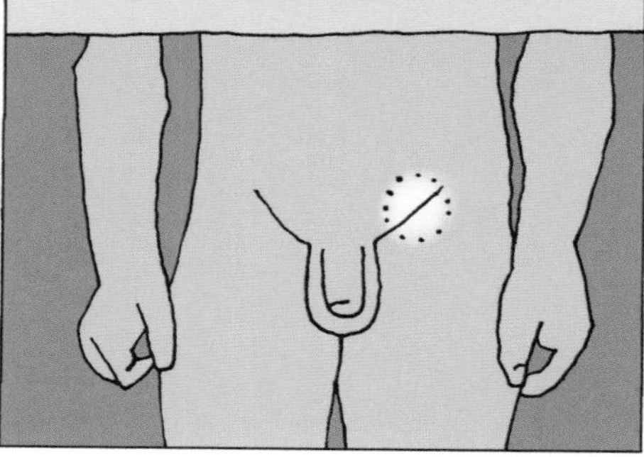
THESE CANALS LEAVE WEAK SPOTS IN THE BODY WALL THAT LEAVE MEN PRONE TO HERNIAS.

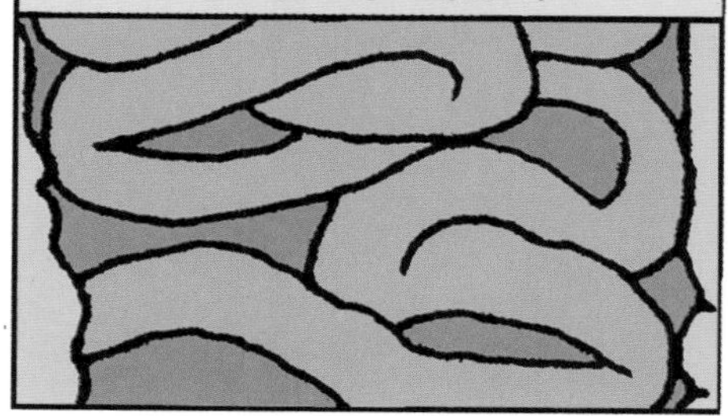
THESE HERNIAS CAN OBSTRUCT THE INTESTINE, AND SOMETIMES CAUSED DEATH IN THE YEARS BEFORE SURGERY.

NO CREATOR WOULD INVENT SUCH A COMPLEX AND DANGEROUS DESIGN.

WE'RE STUCK WITH THIS DESIGN, BECAUSE WE INHERITED OUR DEVELOPMENT FROM FISH-LIKE ANCESTORS.
GLOOP!

THE TESTICULAR DESCENT EVOLVED MUCH LATER AS A CLUMSY ADD-ON.

EACH SPECIES IS BUILT ON OLDER DESIGNS. THE SLATE IS NOT WIPED CLEAN EACH TIME.

NEW PARTS EVOLVE FROM OLD ONES, AND THEY HAVE TO WORK WITH PARTS THAT ARE ALREADY THERE.

THIS CAN LEAD TO SOME STRANGE DEVELOPMENTAL FEATURES.

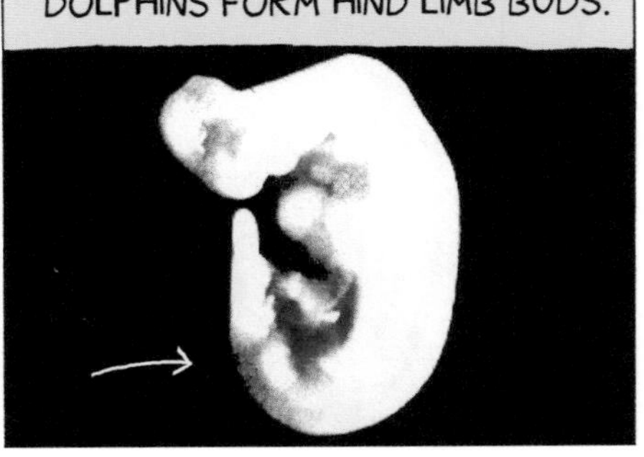
EMBRYONIC WHALES AND DOLPHINS FORM HIND LIMB BUDS.

BULGES OF TISSUE THAT IN A LAND-BASED MAMMAL BECOME THE REAR LEGS.

BUT IN THE MARINE MAMMALS THESE BUDS ARE REABSORBED SOON AFTER THEY'RE FORMED.

BALEEN WHALES, THAT LACK TEETH, BUT WHOSE ANCESTORS WERE TOOTHED WHALES...

DEVELOP EMBRYONIC TEETH THAT VANISH BEFORE BIRTH.

BUT WHAT MAKES THINGS EVOLVE? HOW CAN ONE THING BECOME ANOTHER?

LIFE IS NOT A FLUID THAT FLOWS AND RESHAPES ITSELF DEPENDING ON CONTEXT.

WELL, IT KIND OF IS. THAT'S EXACTLY HOW YOU SHOULD LOOK AT IT.

AND IT IS THE PRIMAL FORCES OF TIME AND GENETICS THAT DO THE MOULDING.

CHECK OUT THESE BUGS DOWN HERE.

HALF ARE GREY AND HALF ARE ORANGE.

THE ORANGE BEETLES ARE MUCH MORE ATTRACTIVE TO PREDATORS...

AND THEREFORE HAVE LESS CHANCE TO REPRODUCE THAN DO THE GREY BEETLES.

THE CAMOUFLAGED GREY BEETLES PASS THE GENETIC COLOUR TRAIT ONTO THEIR OFFSPRING...

SO THAT EVENTUALLY ALL THE INDIVIDUALS IN THE POPULATION WILL BE GREY.

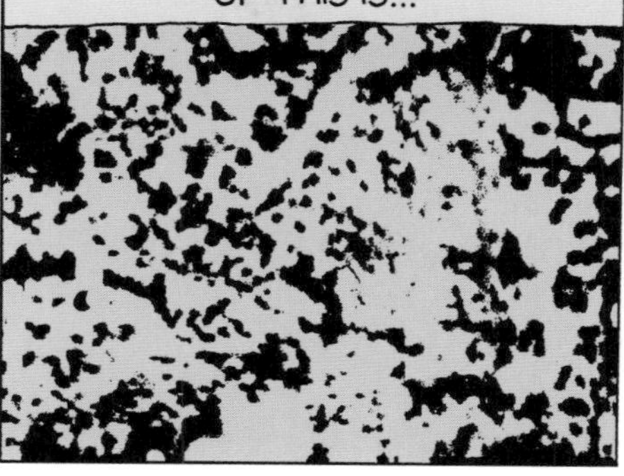
THIS IS NATURAL SELECTION AT WORK. A STRIKING EXAMPLE OF THIS IS...

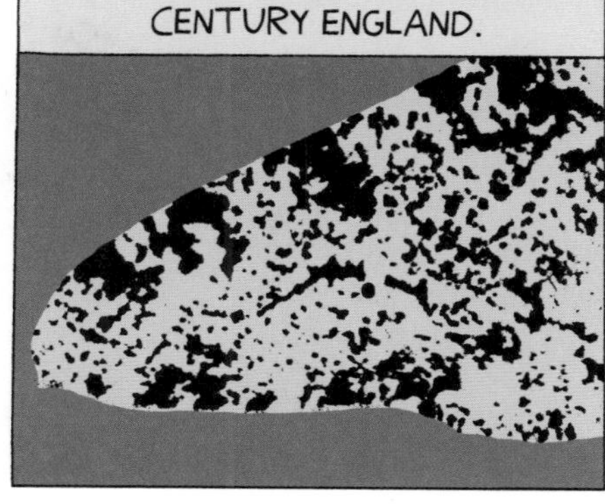
THAT OF THE POPULATION OF DARK MOTHS IN 19TH-CENTURY ENGLAND.

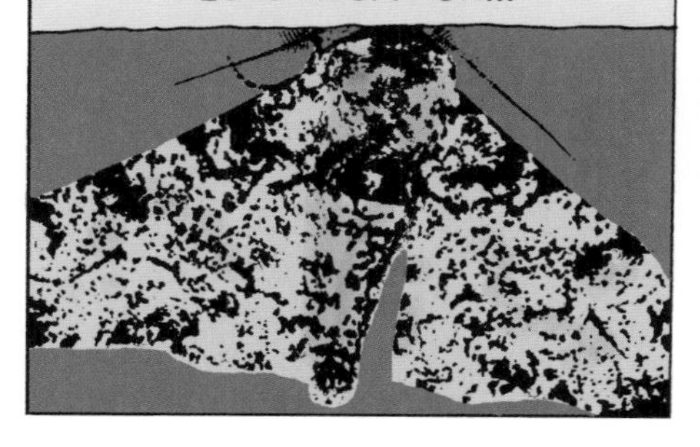
ORIGINALLY THE VAST MAJORITY OF PEPPERED MOTHS HAD LIGHT COLOURISATION...

WHICH CAMOUFLAGED THEM AGAINST THE LIGHT-COLOURED TREES AND LICHENS WHICH THEY RESTED ON.

HOWEVER, BECAUSE OF WIDESPREAD POLLUTION DURING THE INDUSTRIAL REVOLUTION...

MANY LICHENS DIED OUT AND THE TREES BECAME BLACKENED WITH SOOT...

CAUSING MOST OF THE LIGHT-COLOURED MOTHS TO DIE OFF FROM PREDATION.

AT THE SAME TIME THE DARK-COLOURED MOTHS FLOURISHED.

SINCE THEN THERE HAVE BEEN IMPROVED ENVIRONMENTAL STANDARDS...

AND SO LIGHT-COLOURED PEPPERED MOTHS AGAIN BECAME COMMON.

OH YEAH! WELL, I HAPPEN TO KNOW THAT THE WHOLE PEPPERED MOTH THING HAS BEEN REVEALED AS A HOAX.

THERE WERE STUDIES OF THIS PHENOMENON BACK IN THE 1950s.

IN MORE RECENT TIMES, BIOLOGISTS HAVE CRITICISED SOME ASPECTS OF THE METHODOLOGY OF THIS RESEARCH.

AS A RESULT, CERTAIN CRITICS HAVE WRONGLY CLAIMED THAT THE PEPPERED MOTH STORY WAS A HOAX.

BUT THE BASIC ELEMENTS OF THE STORY ARE CORRECT. THE POPULATION OF DARK MOTHS ROSE AND FELL IN PARALLEL WITH INDUSTRIAL POLLUTION.

OF THE SEVERAL FACTORS KNOWN TO PRODUCE EVOLUTIONARY CHANGE...

ONLY NATURAL SELECTION IS CONSISTENT WITH THE PATTERN OF CHANGE SEEN IN THE MOTH POPULATION.

LET ME EXPLAIN. DIFFERENT TYPES OF EYES HAVE EMERGED IN EVOLUTIONARY HISTORY.

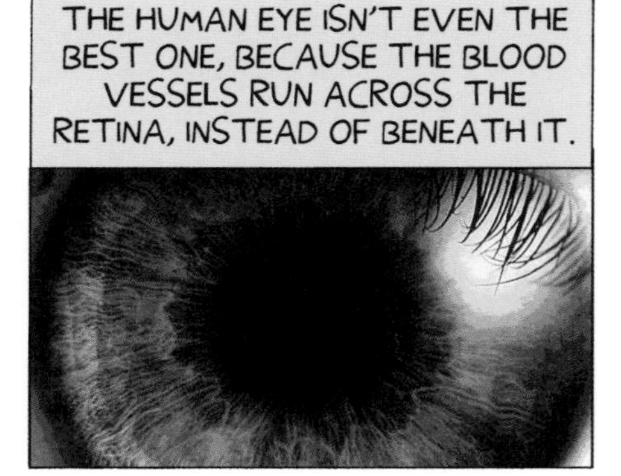

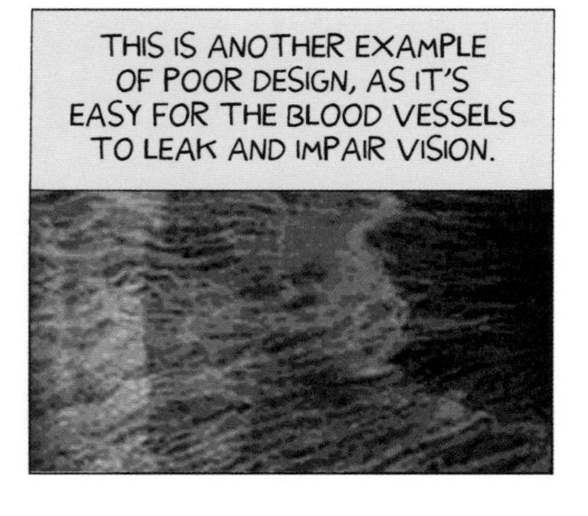

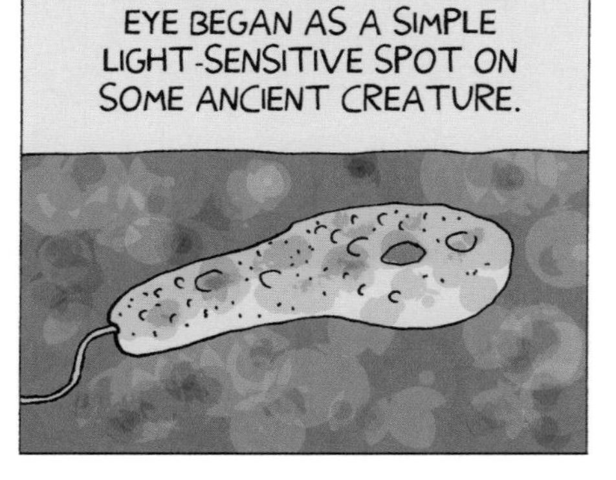

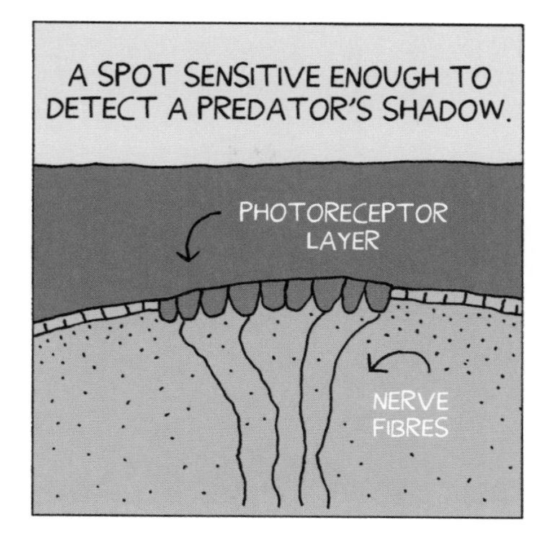
A SPOT SENSITIVE ENOUGH TO DETECT A PREDATOR'S SHADOW.
PHOTORECEPTOR LAYER
NERVE FIBRES

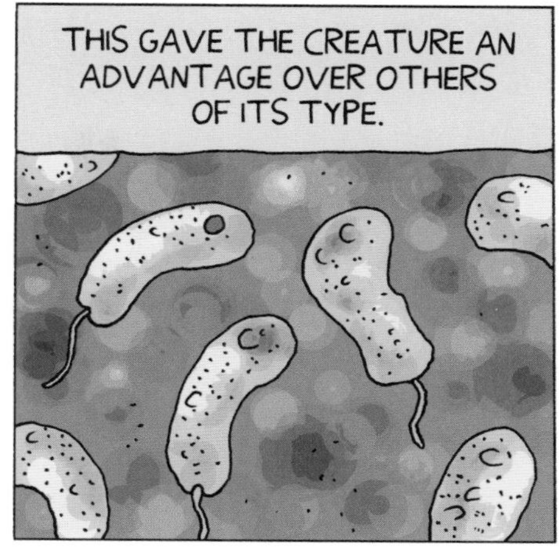
THIS GAVE THE CREATURE AN ADVANTAGE OVER OTHERS OF ITS TYPE.

ITS COMPETITORS DIED, WHILE IT THRIVED, PASSING ITS LIGHT-SENSITIVE SPOT ON TO ITS DESCENDANTS.

EVENTUALLY A RANDOM MUTATION CREATED A DEPRESSION IN THE LIGHT-SENSITIVE PATCH.

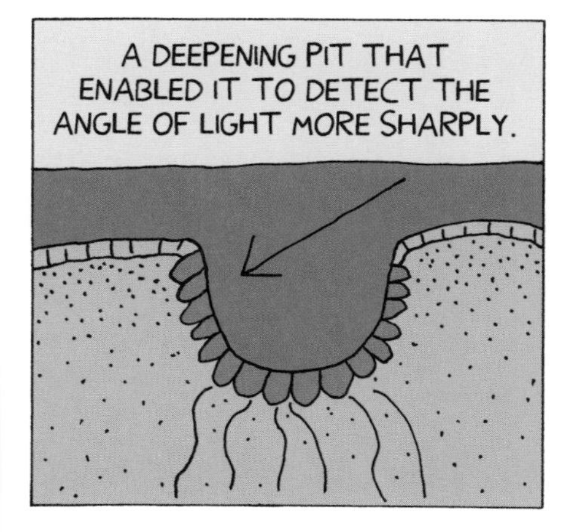
A DEEPENING PIT THAT ENABLED IT TO DETECT THE ANGLE OF LIGHT MORE SHARPLY.

AT EVERY POINT OF ADVANTAGE THE CREATURE PASSED THESE IMPROVEMENTS ONTO ITS OFFSPRING.

AFTER MANY GENERATIONS THE PIT'S OPENING NARROWED, SO THAT LIGHT ENTERED THROUGH A SMALL APERTURE...

MUCH LIKE A PINHOLE CAMERA, ALLOWING THE EYE TO PRODUCE A RUDIMENTARY IMAGE.
OPTIC NERVE

OVER TIME, A LENS FORMED AT THE FRONT OF THE EYE AND CLEAR FLUID FILLED THE CHAMBER...

WHILE THE LIGHT-SENSITIVE PATCH BECAME A RETINA.

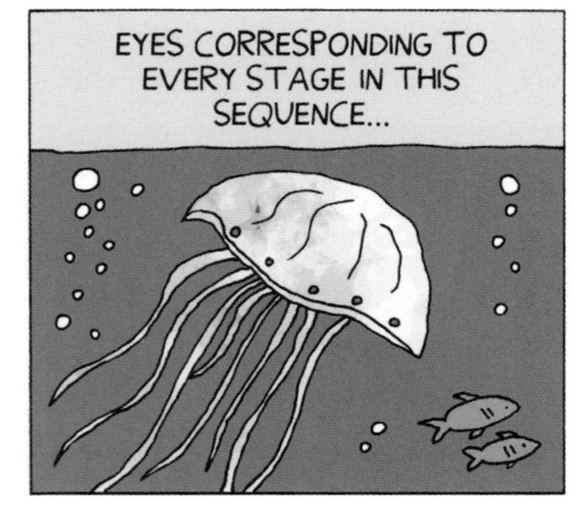
EYES CORRESPONDING TO EVERY STAGE IN THIS SEQUENCE...

HAVE BEEN FOUND IN EXISTING LIVING SPECIES.

BUT WHAT MAKES THESE CHANGES HAPPEN IN THE FIRST PLACE?

FOR THE ANSWER TO THAT WE HAVE TO LOOK DEEP INTO THE MECHANISMS OF LIFE ITSELF.

HOW IT REPLICATES AND THE MISTAKES THAT CAN HAPPEN WHEN IT DOES.

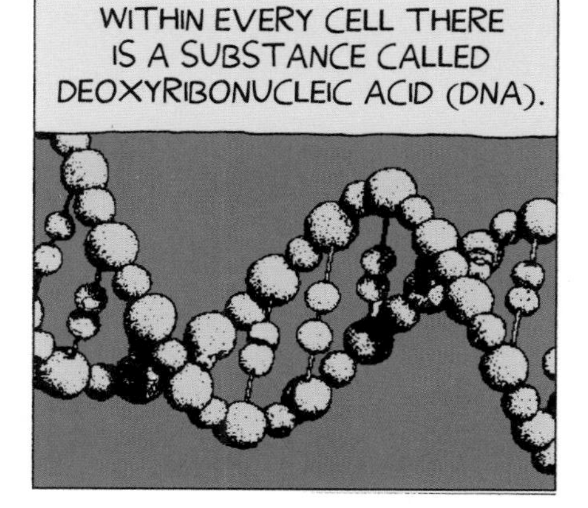
WITHIN EVERY CELL THERE IS A SUBSTANCE CALLED DEOXYRIBONUCLEIC ACID (DNA).

DNA IS VERY FINE AND TIGHTLY COILED. THERE MAY BE AS MUCH AS A METRE IN A SINGLE CELL.

DNA IS REALLY A CODE. IT'S DIVIDED UP INTO SECTIONS CALLED GENES.

YOU HAVE TO THINK OF DNA AS A VAST CHEMICAL INFORMATION DATABASE...

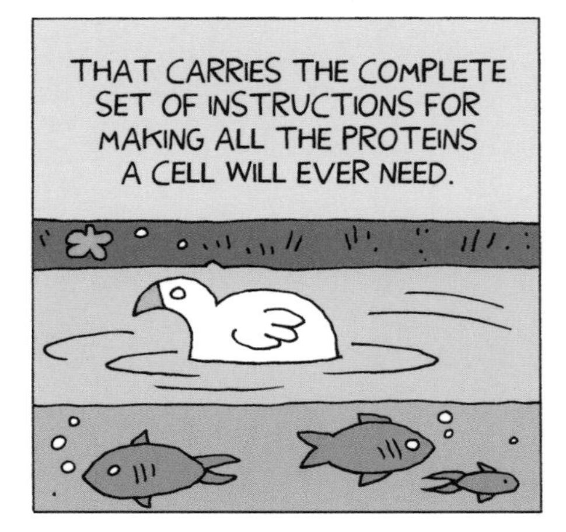
THAT CARRIES THE COMPLETE SET OF INSTRUCTIONS FOR MAKING ALL THE PROTEINS A CELL WILL EVER NEED.

THIS GENETIC INFORMATION IS NOT SO MUCH A BLUEPRINT FOR BUILDING AN ORGANISM...

AS A RECIPE, OUT OF WHICH COMPLEX STRUCTURES CAN THEN EMERGE.

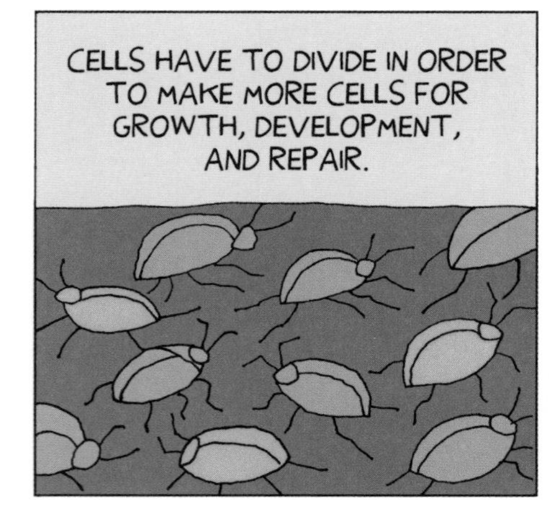
CELLS HAVE TO DIVIDE IN ORDER TO MAKE MORE CELLS FOR GROWTH, DEVELOPMENT, AND REPAIR.

DURING THESE DIVISIONS, MISTAKES IN THE COPYING OF THE DNA STRANDS SOMETIMES HAPPEN.

APART FROM THE COPYING ERRORS THAT OCCUR DURING REPLICATION...

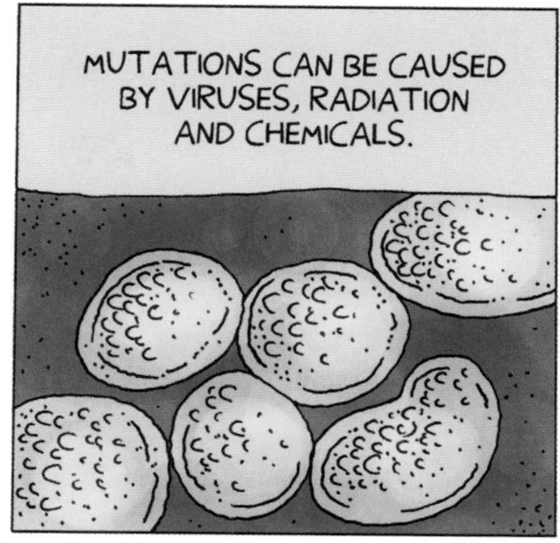
MUTATIONS CAN BE CAUSED BY VIRUSES, RADIATION AND CHEMICALS.

WHEN MUTATIONS HAPPEN THEY CAN HAVE NO EFFECT, THEY CAN BE DAMAGING...

OR THEY CAN OFFER SOME SLIGHT ADVANTAGE IN THE STRUGGLE FOR LIFE.

LIKE THE APPEARANCE OF THE LIGHT-SENSITIVE SPOT ON THE ANCIENT CREATURE PREVIOUSLY DISCUSSED.

IT'S A RANDOM PROCESS, MUCH LIKE THE SHUFFLING OF CARDS.

SOMETIMES AN ORGANISM CAN BE DEALT A USEFUL HAND, AND THEN BECAUSE OF NATURAL SELECTION THESE USEFUL GENETIC TRAITS BUILD UP OVER TIME...

HAS BEEN MORE THAN ENOUGH TIME...

WE SEE THIS IN THE WAY BACTERIA HAVE BECOME RESISTANT TO ANTIBIOTICS.

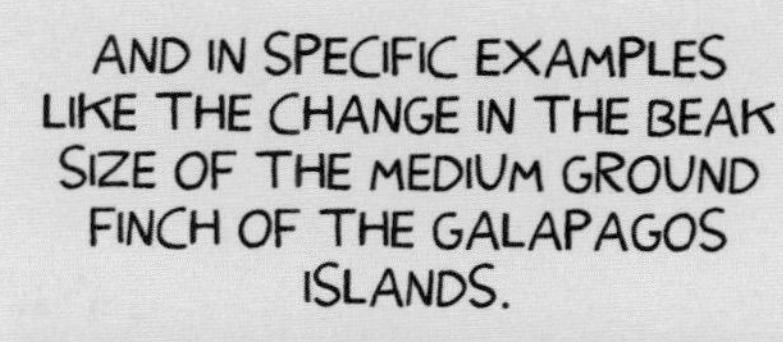
AND IN SPECIFIC EXAMPLES LIKE THE CHANGE IN THE BEAK SIZE OF THE MEDIUM GROUND FINCH OF THE GALAPAGOS ISLANDS.

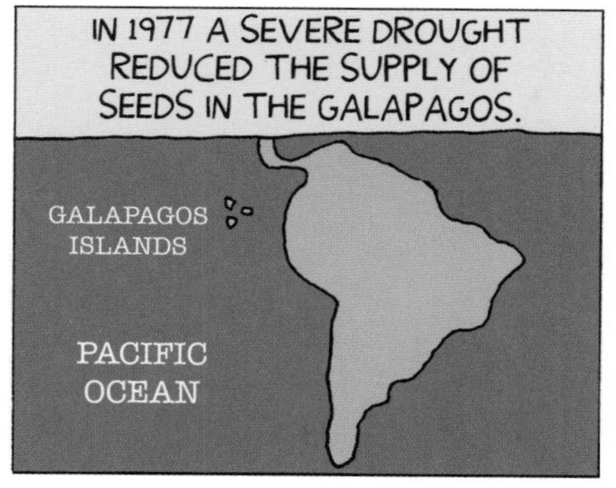
IN 1977 A SEVERE DROUGHT REDUCED THE SUPPLY OF SEEDS IN THE GALAPAGOS.
GALAPAGOS ISLANDS
PACIFIC OCEAN

THIS FINCH, WHICH NORMALLY PREFERRED SMALL SOFT SEEDS...

WAS FORCED TO TURN TO LARGER AND HARDER SEEDS.
?

AS A RESULT, ONLY BIG-BEAKED INDIVIDUALS GOT ADEQUATE FOOD.

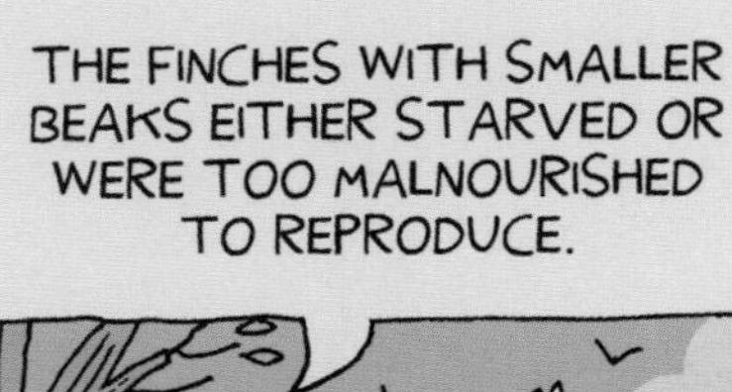

BY THE NEXT GENERATION NATURAL SELECTION HAD INCREASED THE SIZE OF THE FINCHES' BEAKS BY TEN PERCENT.

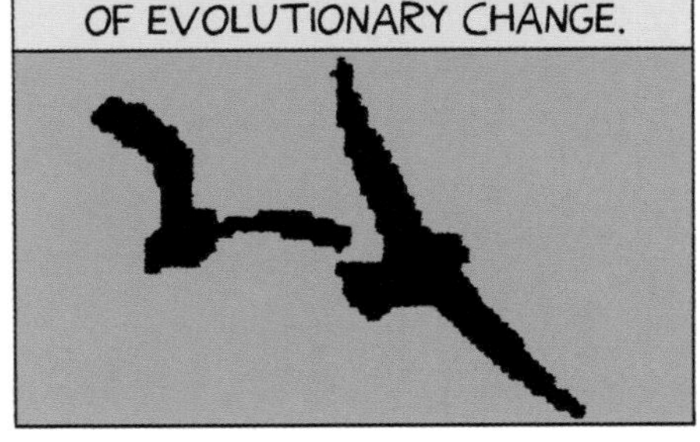

SCIENCE
DENIAL

THE AREAS OF SCIENCE THAT GENERATE THE FIERCEST DEBATE...

TEND TO BE THOSE WE'RE OBLIGED TO TAKE ON TRUST.

THERE IS LITTLE ARGUMENT ANYMORE OVER THE SHAPE OF THE EARTH...

OR THE ROLE OF MICRO-ORGANISMS IN DISEASE.
CHOW!

BUT MORE DIFFICULT CONCEPTS, SUCH AS QUANTUM MECHANICS, NEED A HIGH LEVEL OF SPECIALIST KNOWLEDGE TO BE PROPERLY UNDERSTOOD...

SO THESE AREAS REMAIN THE DOMAIN OF SCIENTISTS.
ER!

HOWEVER, MANY OF SCIENCE'S CONCLUSIONS APPEAR TO CONTRADICT COMMON BELIEF...

AND TO THREATEN IMPORTANT ASPECTS OF OUR LIVES.

HIV DOESN'T CAUSE AIDS. IT'S A BIG PHARMA CONSPIRACY.

THE WORLD WAS CREATED ONLY SIX THOUSAND YEARS AGO. EVERYONE KNOWS THAT.

CLIMATE CHANGE ISN'T HAPPENING, AND EVEN IF IT IS, IT'S NOT CAUSED BY HUMAN ACTIVITY.

SMOKING DOESN'T CAUSE CANCER. MY GRANDMOTHER LIVED UNTIL SHE WAS 103, AND SMOKED SINCE THE DAY SHE WAS BORN.

MY KID GOT AUTISM AFTER GETTING HIS SHOTS. THE VACCINE MUST HAVE DONE IT.

IT SEEMS THAT AN EMOTIONAL STORY COMBINED WITH AN INVESTMENT IN A BELIEF...

CAN TRUMP ANY NUMBER OF SCIENTIFIC STUDIES.

THIS CAN HAVE LETHAL CONSEQUENCES.

THABO MBEKI WAS PRESIDENT OF SOUTH AFRICA FOR NINE AND A HALF YEARS.

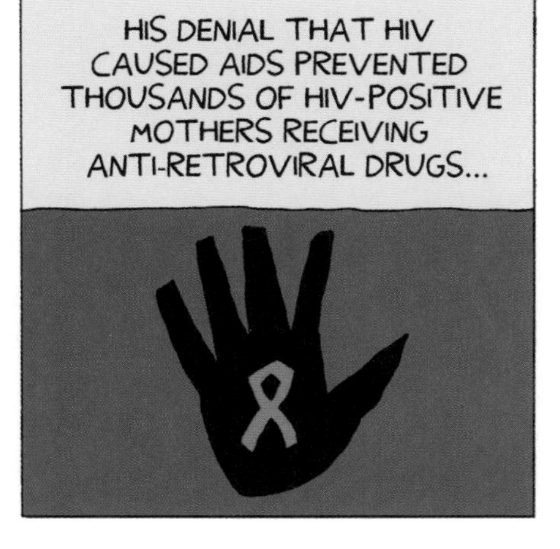
HIS DENIAL THAT HIV CAUSED AIDS PREVENTED THOUSANDS OF HIV-POSITIVE MOTHERS RECEIVING ANTI-RETROVIRAL DRUGS...

AS A RESULT OF WHICH THEY UNNECESSARILY TRANSMITTED THE VIRUS TO THEIR CHILDREN.

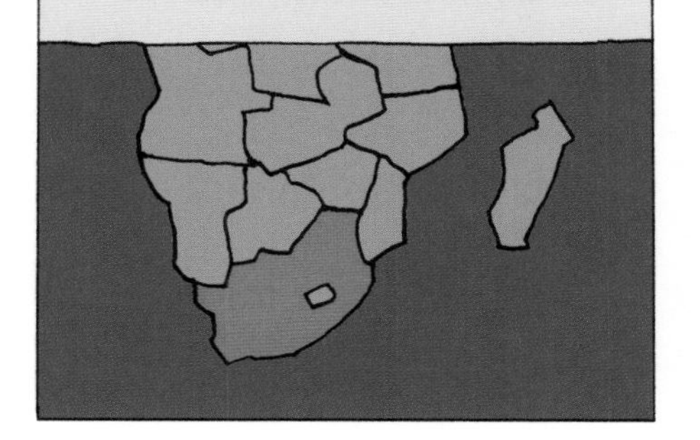

MORE THAN A THIRD OF A MILLION PEOPLE DIED AS A RESULT OF MBEKI'S POLICIES.

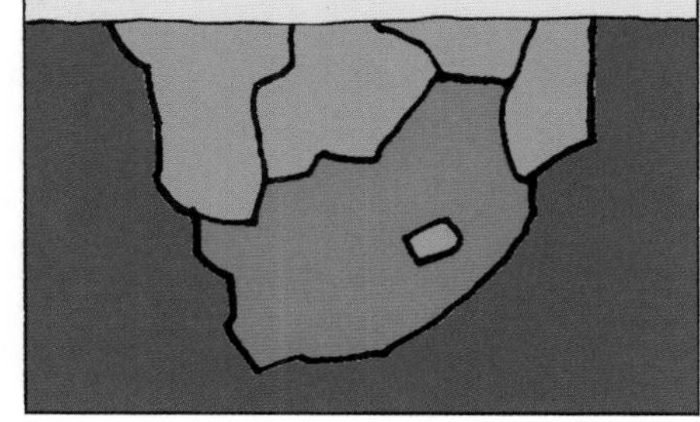

IN 2005, THERE WERE ABOUT 900 DEATHS A DAY.

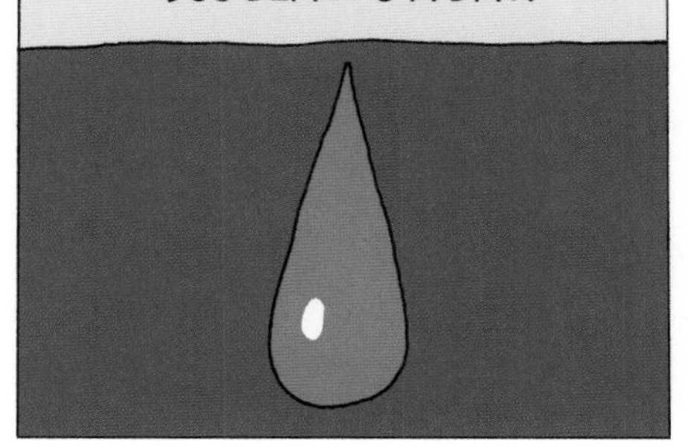

MBEKI'S MINISTER OF HEALTH, MANTO TSHABALALA-MSIMANG...

FAMOUSLY ADVOCATED TREATMENT OF AIDS WITH GARLIC, BEETROOT AND AFRICAN POTATO.

IN CONTRAST BOTH BOTSWANA AND NAMIBIA ACHIEVED GREAT RESULTS WITH THEIR PREVENTING MOTHER-TO-CHILD TRANSMISSIONS PROGRAMMES.

THE HUMAN MIND IS NOTABLE FOR ITS ABILITY TO CLING TO ITS BELIEFS LONG PAST THE POINT...

WHERE ANY EVIDENCE EXISTS TO SUPPORT THOSE BELIEFS.
A FEW SMOKES NEVER HURT ANYONE.

CORPORATE BUSINESS HAS BECOME EXPERT AT EXPLOITING THIS WEAKNESS IN HUMAN PSYCHOLOGY.
YOU HAVE CANCER.

THE STRATEGY OF CREATING DOUBT OVER THE VALIDITY OF SCIENTIFIC RESEARCH...

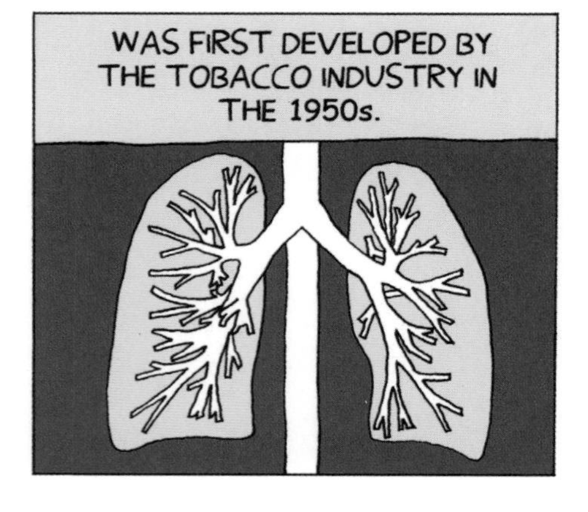
WAS FIRST DEVELOPED BY THE TOBACCO INDUSTRY IN THE 1950s.

FOR OBVIOUS REASONS, IT WAS KEEN TO DISCREDIT THE SCIENTIFIC LINK BETWEEN SMOKING AND ITS HEALTH RISKS.

IT ACHIEVED THIS AIM BY FUNDING RESEARCH THAT SUGGESTED EXPLANATIONS FOR CANCER OTHER THAN SMOKING.
JOE SMOKER RIP

VAST SUMS OF MONEY WERE SPENT ON THIS CAUSE, A FIGURE THAT EXCEEDED $100 MILLION BY THE 1970S.

ANY RESEARCH THAT FAVOURED ITS VIEWPOINT WAS PROMOTED, WHILE RESEARCH THAT DIDN'T WAS SUPPRESSED.

A WHOLE SERIES OF DISSENTING EXPERTS WERE PARADED IN ORDER TO BOLSTER THE ARGUMENT.

IN THIS WAY, TOBACCO FIRMS SUCCESSFULLY MUDDIED THE WATERS OF SCIENTIFIC RESEARCH...

WHILE GIVING THEMSELVES LEGAL MEANS BY WHICH TO OPPOSE REGULATION AND FIGHT COMPENSATION CLAIMS.

THESE TECHNIQUES HAVE BEEN ENTHUSIASTICALLY EMBRACED BY THE OIL AND GAS INDUSTRIES IN ORDER TO THROW DOUBT ON THE REALITY OF CLIMATE CHANGE.

THIS METHOD FAILS TO MAKE A DISTINCTION BETWEEN EVIDENCE AND OPINION...
LET'S HAVE A HEATED DEBATE.

AND GIVES FREE PUBLICITY TO MARGINAL BELIEFS...
GRR!
GRR!

SUCH AS ANTI-VACCINATIONISTS, HOMEOPATHY PROPONENTS...
BLAH BLAH BLAH!

AND CLIMATE CHANGE DENIERS.

MORE THAN NINETY PERCENT OF CLIMATE SCIENTISTS CONSIDER MAN-MADE CLIMATE CHANGE TO BE A REALITY.

HOWEVER, YOU WOULDN'T KNOW THIS FROM THE WAY THE ISSUE HAS BEEN PORTRAYED IN THE MEDIA.

THE MEDIA'S ATTEMPTS AT BALANCE HAVE FAR TOO OFTEN DISTORTED SCIENTIFIC DEBATE...

AND THIS HAS HAD THE EFFECT OF MISLEADING THE GENERAL PUBLIC...
IT'S SNOWING!

INTO THINKING THAT THERE ARE BASIC DIVISIONS AMONG SCIENTISTS WHEN THERE REALLY AREN'T.
SO MUCH FOR CLIMATE CHANGE.

JUST BECAUSE IT'S SNOWING NOW DOESN'T MEAN THE EARTH ISN'T WARMING UP.

ADVERSARIAL DISPUTE MIGHT BE AN EFFECTIVE WAY TO COVER POLITICS...
OH YEAH!

BUT IT DOESN'T HELP AT ALL WHEN IT COMES TO EXPLAINING SCIENTIFIC ISSUES.

HUMANS ARE CHANGING THE CLIMATE. DENIERS DISAGREE, OF COURSE.

BUT THAT'S BECAUSE THEY INSIST ON HOLDING FIXED POSITIONS THAT HAVE NOTHING TO DO WITH SCIENCE.

THE SCIENTIFIC METHOD IS SELF-CORRECTING.
OBSERVATION.

THIS SELF-CORRECTION MAY TAKE TIME...
QUESTIONS.
?

AND ANY RESULTS MAY BE MIRED IN CONTROVERSY UNTIL THE ISSUE IS SETTLED.
RESEARCH.

ONLY SCIENCE CAN REVEAL THE TRUE NATURE OF THE WORLD.
EXPERIMENT.

A SCIENTIST CAN'T AFFORD TO CHERRY-PICK DATA OR IGNORE INCONVENIENT FACTS...
ANALYSE DATA.

BECAUSE ANY RESULTS WILL BE SUBJECT TO THE SAVAGE SCRUTINY OF OTHER SCIENTISTS...
CONCLUSION.
I'VE GOT IT!

WHO WILL BE ONLY TOO KEEN TO PICK HOLES IN ANY THEORY.
THAT, THEN, IS MY CONCLUSION.

YOUR THEORY THAT THE MOON IS HELD IN PLACE BY AN INVISIBLE STRING IS QUITE ABSURD.

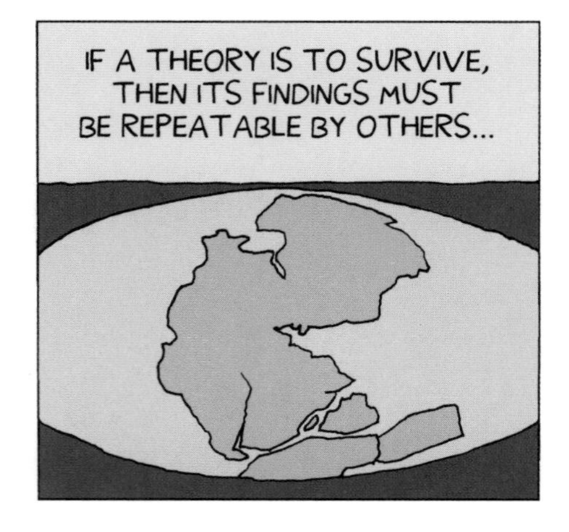
IF A THEORY IS TO SURVIVE, THEN ITS FINDINGS MUST BE REPEATABLE BY OTHERS...

AND IT HAS TO BE ABLE TO BOTH EXPLAIN AND PREDICT EVENTS IN NATURE.

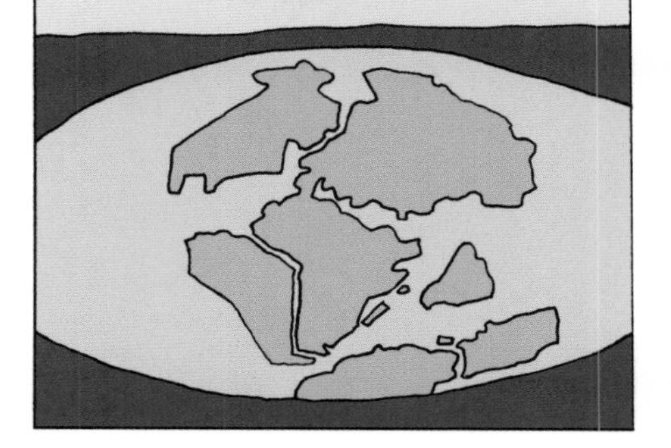

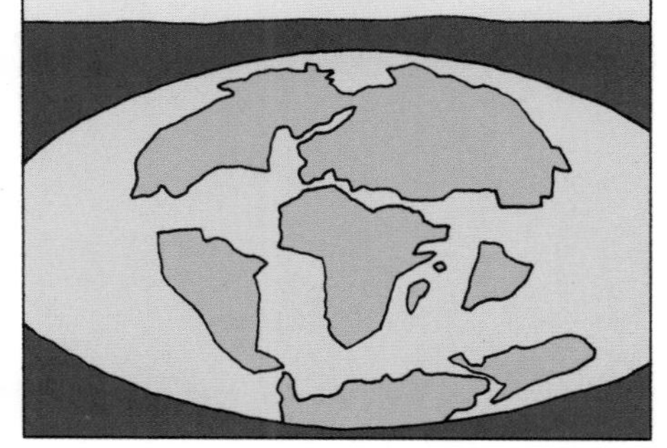

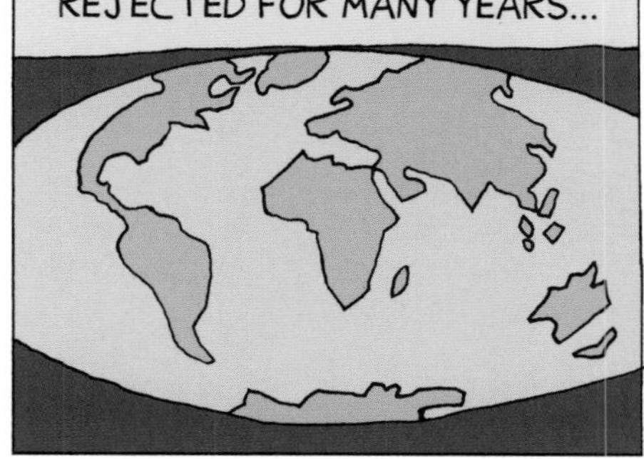

UNTIL THE THEORY OF TECTONIC
PLATES, IN THE 1960s, GAVE AN
EXPLANATION OF HOW SUCH
MOVEMENT WAS POSSIBLE.

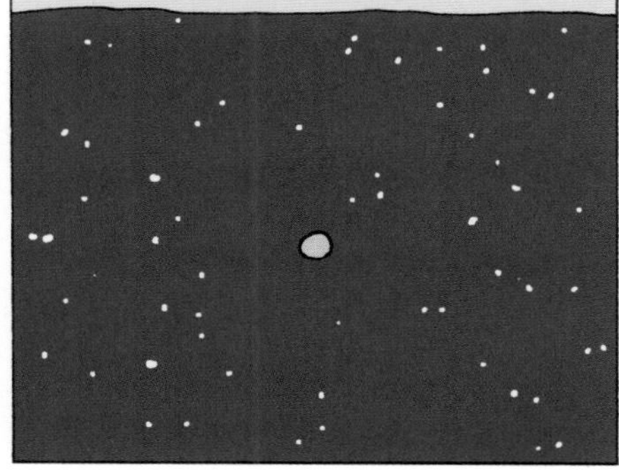

ASTRONOMER FRED HOYLE'S STEADY STATE UNIVERSE THEORY IS A GOOD EXAMPLE OF THIS. HOYLE BELIEVED THAT MATTER WAS CONSTANTLY BEING CREATED BETWEEN THE GALAXIES, MAKING THE UNIVERSE EXPAND FOREVER.

AN INFINITE UNIVERSE, ENDLESS AND ETERNAL, THAT REQUIRED NO CAUSE AND NO CREATOR. AN ELEGANT THEORY, DESTROYED BY ANNOYING FACTS.

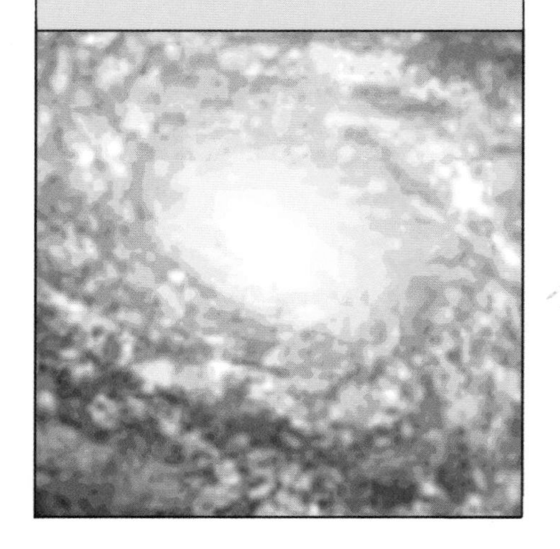
HOYLE'S THEORY WAS FATALLY DAMAGED BY THE DISCOVERY OF COSMIC BACKGROUND RADIATION. THE LAST ECHOES OF THE BIG BANG, PROOF OF A COSMIC STARTING POINT.

SCIENCE HAS A VERY ROBUST SYSTEM FOR ASSESSING THE QUALITY OF RESEARCH BEFORE IT'S PUBLISHED.
EXCUSE ME, BUT I NEED TO TAKE YOUR PAPER FOR REVIEW.

THIS SYSTEM IS CALLED PEER REVIEW.
NO! GO AWAY!

PEER REVIEW MEANS THAT OTHER EXPERTS IN THE SAME FIELD WILL CHECK RESEARCH PAPERS...
WE HAVE TO CHECK IT.

FOR VALIDITY, SIGNIFICANCE AND ORIGINALITY.
IT'S LIKE RUNNING A GAUNTLET.

EDITORS OF JOURNALS DRAW ON A LARGE POOL OF EXPERTS TO SCRUTINISE PAPERS...
DONT WORRY. WE'LL BE GENTLE.

BEFORE DECIDING WHETHER TO PUBLISH THEM.
OR RATHER WE WON'T. HAR!

MANY OF THE RESEARCH CLAIMS THAT APPEAR IN NEWSPAPERS, MAGAZINES AND OTHER MEDIA...
AND NOW THIS.

ARE NOT PUBLISHED IN PEER-REVIEWED JOURNALS.
ELECTRICITY PYLONS ARE DANGEROUS.

IT'S WHY MANY REPORTED FINDINGS, SUCH AS CLAIMS ABOUT WONDER CURES OR NEW HEALTH SCARES...

NEVER COME TO ANYTHING.
I NEVER HURT ANYONE.

AS A SOCIETY WE CAN'T BASE PUBLIC POLICY ON WORK THAT HASN'T BEEN ASSESSED AND COULD BE FLAWED.

PUBLICATION IN A PEER-REVIEWED JOURNAL IS ONLY THE FIRST STEP.

ANY FINDINGS AND THEORIES MUST GO ON TO BE RE-TESTED AGAINST OTHER WORK IN THE SAME FIELD.
MY PAPER'S BEEN ACCEPTED BY THIS JOURNAL.

SOME PAPERS' CONCLUSIONS WILL BE DISPUTED...
NOW THE SCRUTINY REALLY BEGINS.

WHILE FURTHER RESEARCH MAY SHOW THAT A PAPER NEEDS REVISION AS MORE DATA IS GATHERED.
BACK TO THE DRAWING BOARD.

THE PEER-REVIEW PROCESS CONNECTS LIKE-MINDED PEOPLE, LETTING THEM KNOW ABOUT NEW RESEARCH IN THEIR FIELD.
A SCIENTIST HAS TO HAVE THICK SKIN. SIGH!

IT'S A PERMANENT RECORD OF WHAT HAS BEEN DISCOVERED AND BY WHOM.
nature
HE LANCET
Science

IT HELPS SCIENTISTS PROMOTE THEIR WORK AND GAIN RECOGNITION FROM FUNDERS AND OTHER INSTITUTIONS.
TAKE THIS NOBEL PRIZE.

IT CAN'T ALWAYS DETECT FRAUD OR MISCONDUCT. IF SOMEONE SETS OUT TO FALSIFY RESULTS, THEN THERE MAY BE NO WAY OF KNOWING THIS...

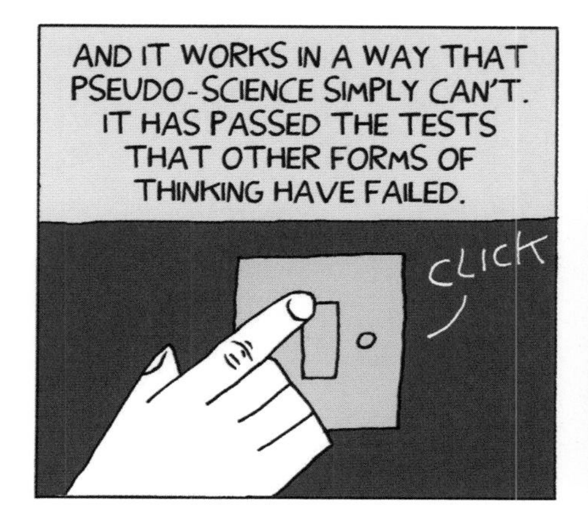
AND IT WORKS IN A WAY THAT PSEUDO-SCIENCE SIMPLY CAN'T. IT HAS PASSED THE TESTS THAT OTHER FORMS OF THINKING HAVE FAILED.
CLICK

SCIENCE HAS BOTH BUILT THE TECHNOLOGICAL WORLD WE LIVE IN...

AND REVEALED TO US MANY OF THE DEEP MYSTERIES OF THE UNIVERSE.
HUH?

SCIENCE BUILDS AND ORGANISES KNOWLEDGE IN THE FORM OF TESTABLE EXPLANATIONS AND PREDICTIONS.

SCIENCE IS THE MOST SUCCESSFUL TOOL EVER DEVISED FOR EXPLAINING OUR UNIVERSE.

THIS IS WHY SCIENCE PROCEEDS THE WAY IT DOES AND WHY IT IS SO POWERFUL.
END

ACKNOWLEDGEMENTS:

This book wouldn't exist without the fine work of the many science writers and journalists whose work I've read in the process of research. So thanks in particular to Phil Plait, Jerry Coyne, Brian Deer, Ben Goldacre, Steven Novella, Simon Singh and Edzard Ernst.

I'd also like to thank the people who have supported me through the long process of this book's creation: Ian Williams, Scott McCloud, Pádraig Ó Méalóid, Peter Clack, Sue Krekorian, Sarah MacIntyre, Megan Donnelly, John Miers, Jonathan Edwards, Louise Evans, Paul Gravett, Peter Stanbury, Joe Gordon, Corinne Pearlman and all at Myriad.

Electroconvulsive Therapy

Adams C. What happens in electroshock therapy? *The Straight Dope*. 1999 March 19. www.straightdope.com/columns/read/1311/what-happens-in-electroshock-therapy (accessed 2011 Oct 24).

Benbow SM. Adverse effects of ECT. In Scott AIF, editor. *The ECT Handbook*. 2nd ed. London: Royal College of Psychiatrists; 2004. p.170–74.

Bell V. ECT: the blues and the electric avenue. *Mind Hacks*. 2008 Jan 5. http://mindhacks.com/2008/01/05/ect-the-blues-and-the-electric-avenue (accessed 2011 Oct 24).

UK ECT Review Group. Efficacy and safety of electroconvulsive therapy in depressive disorders: a systematic review and meta-analysis. *The Lancet* 2003 Mar. p.799–808.

Rose D, Fleischmann P, Wykes T, *et al*. Patients' perspectives on electroconvulsive therapy. *BMJ* 2003;326. 2003 June 19. www.bmj.com/content/326/7403/1363.full (accessed 2011 Oct 24).

Sackeim HA, Prudic J, Fuller R, *et al*. The cognitive effects of electroconvulsive therapy in community settings. *Neuropsychopharmacology* 2007 Jan. p.244–54.

Homeopathy

What is Homeopathy? *10:23*. www.1023.org.uk/what-is-homeopathy.php (accessed 2011 Oct 24).

Singh S. Homeopathy; what's the harm? *10:23*. www.1023.org.uk/whats-the-harm-in-homeopathy.php (accessed 2011 Oct 24).

Singh S. Could water really have a memory? *BBC News*. 2008 July 25. http://news.bbc.co.uk/1/hi/health/7505286.stm (accessed 2011 Oct 24).

Richmond C. Obituary: Jacques Benveniste; Maverick scientist behind a controversial experiment into the efficacy of homeopathy. *Guardian*, 2004 Oct 21. www.guardian.co.uk/science/2004/oct/21/obituaries.guardianobituaries (accessed 2011 Oct 24).

Goldacre B. Newsnight/Sense about science malaria & homeopathy sting – the transcripts. *Bad Science*. 2006 Sep 1. www.badscience.net/2006/09/newsnightsense-about-science-malaria-homeopathy-sting-the-transcripts.

Hope A. State Coroner. Coronial inquest into the death of Penelope Dingle. www.safetyandquality.health.wa.gov.au/docs/mortality_review/inquest_finding/Dingle_Finding.pdf (accessed 2011 Oct 24).

The Facts in the Case of Dr. Andrew Wakefield

Gorski, D. The fall of Andrew Wakefield. *Science-Based Medicine*. 2010 Feb 22. www.sciencebasedmedicine.org/index.php/the-fall-of-andrew-wakefield (accessed 2011 Oct 24).

Deer B. The *Lancet* scandal. http://briandeer.com/mmr-lancet.htm (accessed 2011 Oct 24).

Deer B. The Wakefield factor. http://briandeer.com/wakefield-deer.htm (accessed 2011 Oct 24).

Deer B. Solved – the riddle of MMR. http://briandeer.com/solved/solved.htm (accessed 2011 Oct 24).

Deer B. MMR doctor given legal aid thousands. *Sunday Times* 2006 Dec 31. http://briandeer.com/mmr/st-dec-2006.htm (accessed 2011 Oct 24).

Deer B. How the case against the MMR vaccine was fixed. *BMJ* 2011; 342:c5347. 2011 Jan 5. www.bmj.com/content/342/bmj.c5347.full (accessed 2011 Oct 24).

Godlee F, Smith J, Marcovitch H. Wakefield's article linking MMR vaccine and autism was fraudulent. *BMJ* 2011; 342:c7452. 2011 Jan 5. www.bmj.com/content/342/bmj.c7452.full (accessed 2011 Oct 24).

Boseley S. Andrew Wakefield found 'irresponsible' by GMC over MMR vaccine scare. *Guardian.* 2010 Jan 28. www.guardian.co.uk/society/2010/jan/28/andrew-wakefield-mmr-vaccine (accessed 2011 Oct 24).

Goldacre G. *Bad Science.* London: Fourth Estate; 2008.

Goldacre G. The media's MMR hoax. *Bad Science.* 2008 Aug 30. www.badscience.net/2008/08/the-medias-mmr-hoax (accessed 2011 Oct 24).

Chiropractic

Singh S, Ernst E. *Trick or Treatment? Alternative Medicine on Trial.* London: Bantam Press; 2008.

Adams C. Is chiropractic for real or just quackery? *The Straight Dope.* 2008 June 2. www.straightdope.com/columns/read/2771/is-chiropractic-for-real-or-just-quackery (accessed 2011 Nov 13).

Singh S. Beware the spinal trap. *Guardian.* 2008 April 19. www.guardian.co.uk/commentisfree/2008/apr/19/controversiesinscience-health (accessed 2011 Nov 13).

Novella S. Chiropractic – a brief overview. *Science-Based Medicine.* 2009 June 24. www.sciencebasedmedicine.org/index.php/chiropractic-a-brief-overview-part-i (accessed 2011 Nov 13).

Wolff J. Deadly twist: Neck adjustments can be risky. *MSNBC.* 2007 June 17. www.msnbc.msn.com/id/18871755/#.Trlwrxzx-fl (accessed 2011 Nov 13).

What alternative health practitioners might not tell you. *ebm-first.* 2011 Oct 5. www.ebm-first.com/chiropractic/risks.html (accessed 2011 Nov 13).

Hansen J. Doctors accuse chiropractors of selling anti-vaccination message. *News.com.au.* 2011 July 27. www.news.com.au/national/doctors-accuse-chiropractors-of-selling-anti-vaccination-message/story-e6frfkvr-1226102836863 (accessed 2011 Nov 13).

The Moon Hoax

Plait P. Fox TV and the Apollo moon hoax. *Bad Astronomy.* 2001 Feb 13. www.badastronomy.com/bad/tv/foxapollo.html (accessed 2011 Oct 24).

Plait P. *Bad Astronomy: Misconceptions and Misuses Revealed, from Astrology to the Moon Landing 'Hoax'.* New York: John Wiley & Sons; 2002.

Phillips A. The great moon hoax. *Nasa Science News.* 2001 Feb 23. http://science.nasa.gov/science-news/science-at-nasa/2001/ast23feb_2 (accessed 2011 Oct 24).

Shermer M. Fox goes to the moon, but NASA never did: the no-moonies cult strike. 2001. http://homepages.wmich.edu/~korista/moonhoax2.html (accessed 2011 Nov 13).

Fracking

Environmental Protection Agency. Hydraulic Fracturing Background Information. http://water.epa.gov/type/groundwater/uic/class2/hydraulicfracturing/wells_hydrowhat.cfm (accessed 2012 June 12).

Pro Publica. What Is Hydraulic Fracturing? www.propublica.org/special/hydraulic-fracturing-national (accessed 2012 June 12).

Michael K. The 10 Scariest Chemicals Used In Hydraulic Fracking. *Business Insider.* 2012 Mar 16. www.businessinsider.com/scary-chemicals-used-in-hydraulic-fracking-2012-3?op=1 (accessed 2012 June 12).

Lustgarten A. Injection Wells: The Poison Beneath Us. Pro Publica. 2012 June 21. www.propublica.org/article/injection-wells-the-poison-beneath-us (accessed 2012 July 2nd).

Urbina I. Regulation Lax as Gas Wells' Tainted Water Hits Rivers. *New York Times*. 2011 Feb 26. www.nytimes.com/2011/02/27/us/27gas.html?_r=3&pagewanted=all (accessed 2012 June 20).

Drajem M, Efstathiou Jr J. Cabot's Methodology Links Tainted Water Wells to Gas Fracking. *Bloomberg News*. 2012 Nov 2. www.bloomberg.com/news/2012-10-02/cabot-s-methodology-links-tainted-water-wells-to-gas-fracking.html (accessed 2012 Nov 21).

The Halliburton Loophole. *New York Times*. 2009 Nov 2. www.nytimes.com/2009/11/03/opinion/03tue3.html?_r=0 (accessed 2012 June 20).

Harvey F. UK fracking should be expanded, but better regulated, says report. *Guardian*. 2012 June 29. http://www.guardian.co.uk/environment/2012/jun/29/shale-gas-fracking-expanded-regulated (accessed 2012 July 20).

Merrick J, Chorley M. Osborne accused over gas lobbyist father-in-law. *Independent*. 2012 J uly 29. www.independent.co.uk/news/uk/politics/osborne-accused-over-gas-lobbyist-fatherin-law-7985001.html?origin=internalSearch (accessed 2012 Aug 5).

Gas Drilling In Balcombe. Balcombe MP appointed Cuadrilla director to government. 2012 Feb 1. http://gasdrillinginbalcombe.wordpress.com/2012/02/01/balcombe-mp-appointed-cuadrilla-director-to-government (accessed 2012 Aug 5)

Climate Change

Carey B. January to be coldest since 1985, where's global warming? *Live Science*. 2011 Jan 07. www.livescience.com/9227-january-coldest-1985-global-warming.html (accessed 2011 Nov 13).

Oliver C, Frigieri G, Clark D. Everything you need to know about climate change – interactive. *Guardian*. 2011 Aug 15. www.guardian.co.uk/environment/interactive/2011/aug/15/everything-know-climate-change (accessed 2011 Nov 13).

Cook J. 10 Indicators of a human fingerprint on climate change. *Skeptical Science*. 2010 July 30. www.skepticalscience.com/10-Indicators-of-a-Human-Fingerprint-on-Climate-Change.html (accessed 2011 Nov 13).

Dow K, Downing TE. *The Atlas of Climate Change*. London: Earthscan; 2006.

Carrington D. Q&A: 'Climategate'. *Guardian*. 2010 July 7. www.guardian.co.uk/environment/2010/jul/07/climate-emails-question-answer (accessed 2011 Nov 13).

Monbiot G. The climate denial industry is out to dupe the public. And it's working. *Guardian*. 2009 Dec 7. www.guardian.co.uk/commentisfree/cif-green/2009/dec/07/climate-change-denial-industry (accessed 2011 Nov 13)

Mayer J. Covert operations: The billionaire brothers who are waging a war against Obama. *New Yorker*. 2010 Aug 30. www.newyorker.com/reporting/2010/08/30/100830fa_fact_mayer?currentPage=all (accessed 2011 Nov 13).

Mandia S. Global Warming: Man or Myth? Climategate coverage: unfair & unbalanced. http://profmandia.wordpress.com/ 2010/04/18/climategate-coverage-unfair-unbalanced (accessed 2011 Oct 24).

Oreskes N, Conway EM. *Merchants of Doubt: How a Handful of Scientists Obscured the Truth on Issues From Tobacco Smoke to Global Warming*. New York: Bloomsbury; 2010.

Evolution

National Academy of Sciences and Institute of Medicine. Is evolution a theory or a fact? http://nationalacademies.org/evolution/TheoryOrFact.html (accessed 2011 Nov 8).

Coyne JA. *Why Evolution is True*. Oxford: Oxford University Press; 2009.

Jones S. *Almost Like a Whale: The Origin of Species Updated.* New York: Ballantine; 1999.
Dawkins R. *The Selfish Gene: 30th Anniversary Edition.* Oxford: Oxford University Press; 2006.

Shubin N. *Your Inner Fish: The amazing discovery of our 375-million-year-old ancestor.* London: Penguin; 2009.

Gould SJ. *Hen's Teeth and Horse's Toes: Further Reflections on Natural History.* London: Penguin; 1990.
Grant S. Fine tuning the peppered moth paradigm. *Evolution* 1999. 53(3), p.980–84.

Weiner J. *The Beak of the Finch: A Story of Evolution in Our Time.* New York: Vintage; 1995.

Science Denial

McKee M. Denialism: what is it and how should scientists respond? *European Journal of Public Health* 2009 Vol. 9, Nos.1, 2–4.

Boseley S. Mbeki Aids denial 'caused 300,000 deaths'. *Guardian.* 2008 Nov 26. www.guardian.co.uk/world/2008/nov/26/aids-south-africa (accessed 2011 Nov 13).

Beresford D. Manto Tshabalala-Msimang obituary. *Guardian.* 2010 Jan 7. www.guardian.co.uk/world/2010/jan/07/manto-tshabalala-msimang-obituary (accessed 2011 Nov 13).

Foucart S. When science is hidden behind a smokescreen. *Guardian.* 2011 June 28. www.guardian.co.uk/science/2011/jun/28/study-science-research-ignorance-foucart? (accessed 2011 Nov 13).

Eriksen M, Mackay J, Shafey O. *The Tobacco Atlas.* 2nd ed. Atlanta: American Cancer Society; 2006.

Goodell J. As the World Burns: How Big Oil and Big Coal mounted one of the most aggressive lobbying campaigns in history to block progress on global warming. *Rolling Stone.* 2010 Jan 6. www.rollingstone.com/politics/news/as-the-world-burns-20100106 (accessed 2011 Nov 13).
Harris W. How the scientific method works. *How Stuff Works.* http://science.howstuffworks.com/innovation/scientific-experiments/scientific-method2.htm (accessed 2011 Nov 13).

McKie R. Fred Hoyle: the scientist whose rudeness cost him a Nobel prize. *Guardian.* 2010 Oct 3. www.guardian.co.uk/science/2010/oct/03/fred-hoyle-nobel-prize?INTCMP=ILCNETTXT3487 (accessed 2011 Nov 13).

McKie R. Science and truth have been cast aside by our desire for controversy. *Guardian.* 2011 July 24. www.guardian.co.uk/commentisfree/2011/jul/24/science-reporting-climate-change-sceptics (accessed 2011 Nov 13).

MORE FROM MYRIAD EDITIONS

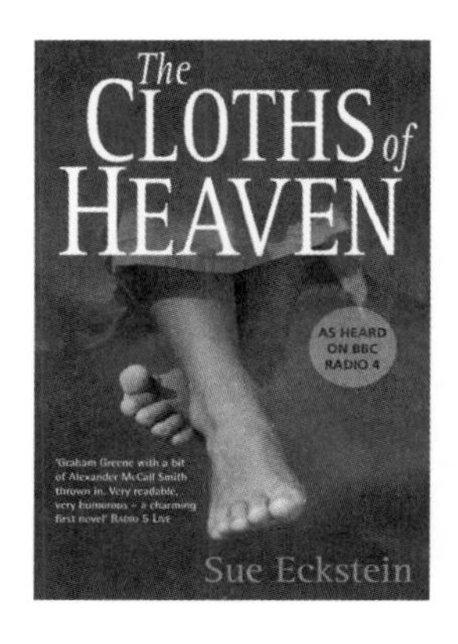

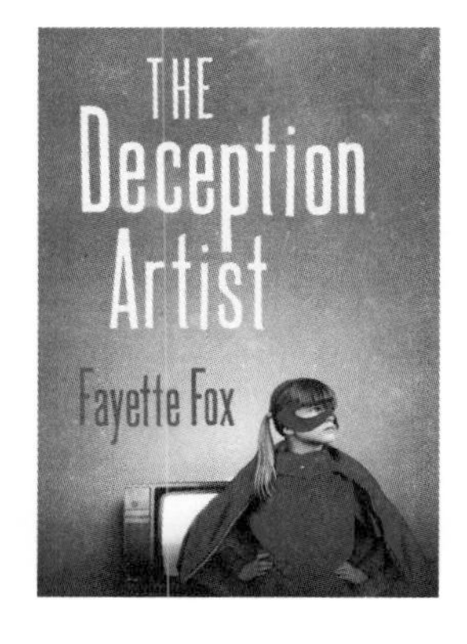

MORE FROM MYRIAD EDITIONS

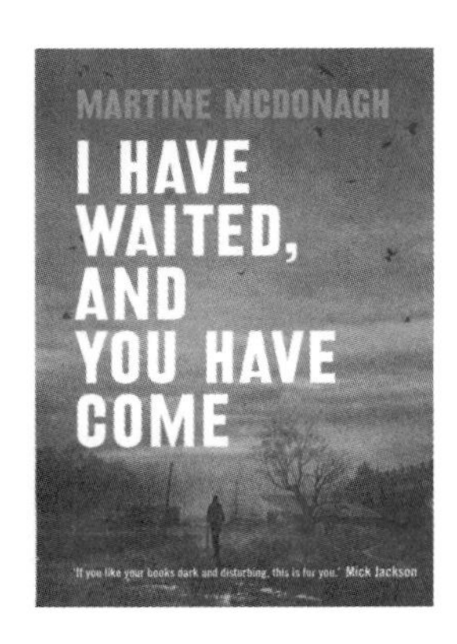

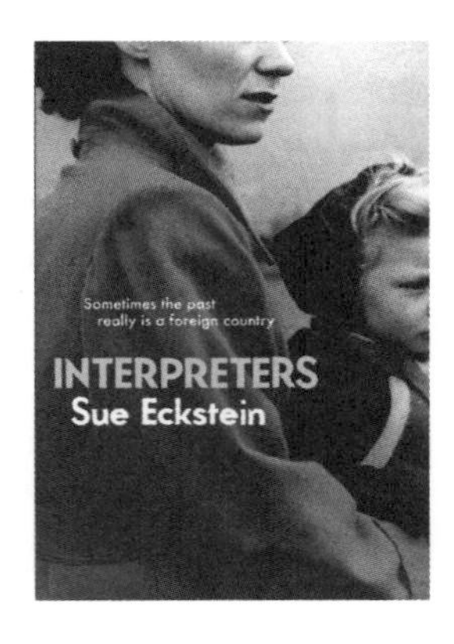

MORE FROM MYRIAD EDITIONS

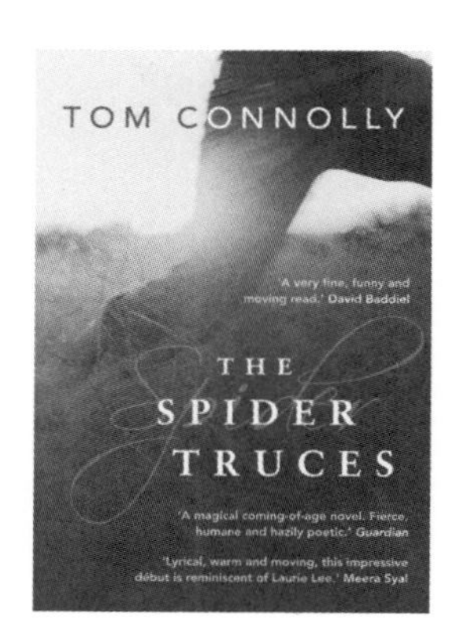

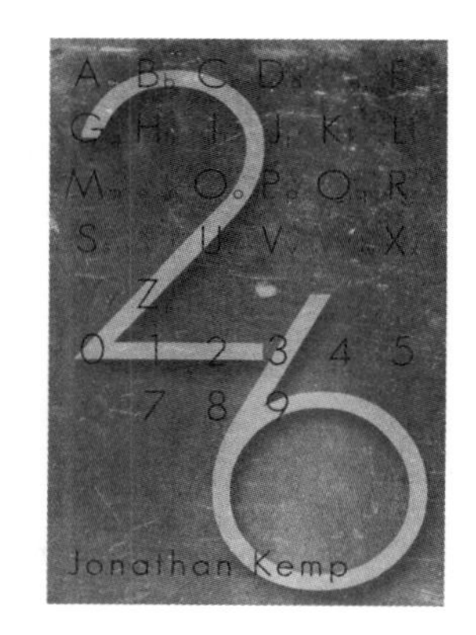

www.myriadeditions.com